PROGRESS IN CLINICAL AND BIOLOGICAL RESEARCH

RECENT TITLES

Please contact publisher for information about previous titles in this series.

MOLECULAR ENDOCRINOLOGY AND STEROID HORMONE ACTION

MOLECULAR ENDOCRINOLOGY AND STEROID HORMONE ACTION

Proceedings of the Fourth International Symposium on
Cellular Endocrinology held in Lake Placid,
New York, August 24–27, 1988

Editors

Gordon H. Sato
James L. Stevens
W. Alton Jones Cell Science Center
Lake Placid, New York

ALAN R. LISS, INC ● NEW YORK

GENETICS  INSTITUTE*

INFORMATION CENTER

CID 1887

Address all Inquiries to the Publisher
Alan R. Liss, Inc., 41 East 11th Street, New York, NY 10003

The publication of this volume was facilitated by the authors and
editors who submitted the text in a form suitable for direct reproduction
without subsequent editing or proofreading by the publisher.

Library of Congress Cataloging-in-Publication Data

International Symposium on Cellular Endocrinology (4th : 1988 : Lake Placid, N.Y.)
   Molecular endocrinology and steroid hormone action : proceedings of the Fourth
International Symposium on Cellular Endocrinology held in Lake Placid, New York,
August 24–27, 1988 / editors, Gordon H. Sato, James L. Stevens.
   p.   cm.—(Progress in clinical and biological research ; v. 322)
ISBN 0-471-56682-9
   1. Steroid hormones—Receptors—Congresses.   2. Molecular endocrinology—
Congresses.   I. Sato, Gordon.   II. Stevens, James L., 1944–   .   III. Title.   IV. Series.
QP572.S7I59   1988   89-13216
612.4'05—dc20   CIP

Contents

Contributors

M. Atger, Hormones et Reproduction, INSERM U. 135, Faculté de Médecine, Paris-Sud, 94275 Le Kremlin-Bicêtre, Cedex France [33]

F. Auricchio, Il Cattedra di Patologia Generale, I Facolta di Medicina e Chirurgia, Universita de Napoli, Italy [133]

A. Bailly, Hormones et Reproduction, INSERM U. 135, Faculté de Médecine, Paris-Sud, 94275 Le Kremlin-Bicêtre, Cedex France [33]

Roberta Binder, Department of Pharmacological Sciences, Health Sciences Center, State University of New York, Stony Brook, NY 11794 [227]

Jack Bodwell, Department of Physiology, Dartmouth Medical School, Hanover, NH 03756 [97]

Jan Carlstedt-Duke, Department of Medical Nutrition, Karolinska Institute, Huddinge University Hospital F69, S-141 86 Huddinge, Sweden [65]

G. Castoria, Il Cattedra di Patologia Generale, Instituto di Patologia Generale, I Facolta di Medicina e Chirurgia, Universita de Napoli, Italy [133]

Chawnshang Chang, Ben May Institute, Department of Surgery/Urology, University of Chicago, Chicago, IL 60637 [53]

C. Chiapetta, Department of Pharmacology, University of Texas Medical School, Houston, TX 77225 [213]

Robert Clarke, Lombardi Cancer Research Center (S128), Georgetown University Medical Center, Washington, DC 20007 [243]

Richard N. Day, Department of Physiology and Biophysics, University of Iowa, Iowa City, IA 52242 [159]

Nooshine Dayani, Baylor College of Medicine, Houston, TX 77030 [83]

Marc Denis, Department of Medical Nutrition, Karolinska Institute, Huddinge University Hospital F69, S-141 86 Huddinge, Sweden [65]

Eugene R. DeSombre, Ben May Institute, University of Chicago, Chicago, IL 60637 [17, 295]

Robert B. Dickson, Lombardi Cancer Research Center (S128), Georgetown University Medical Center, Washington, DC 20007 [243]

M. Di Domenico, Il Cattedra di Patologia Generale, Istituto di Patologia Generale, I Facolta di Medicina e Chirurgia, Universita de Napoli, Italy [133]

Yu Dong, Department of Medical Nutrition, Karolinska Institute, Huddinge University Hospital F69, S-141 86 Huddinge, Sweden [65]

R.M. Gardner, Department of Pharmacology, University of Texas Medical School, Houston, TX 77225 [213]

S. John Gatley, Department of Radiology, University of Chicago, Chicago, IL 60637 [295]

G. Georgiade, Department of Surgery, Duke University Medical Center, Durham, NC 27710 [279]

David A. Gordon, Department of Pharmacological Sciences, Health Sciences Center, State University of New York, Stony Brook, NY 11794 [227]

A. Guiochon-Mantel, Hormones et Reproduction, INSERM U. 135, Faculté de Médecine, Paris-Sud, 94275 Le Kremlin-Bicêtre, Cedex France [33]

Jan-Åke Gustafsson, Department of Medical Nutrition, Karolinska Institute, Huddinge University Hospital F69, S-141 86 Huddinge, Sweden [65]

Paul V. Harper, Department of Radiology, University of Chicago, Chicago, IL 60637 [295]

Kathryn B. Horwitz, Department of Medicine, University of Colorado, Health Sciences Center, Denver, CO 80262 [41]

Alun Hughes, Ben May Institute, University of Chicago, Chicago, IL 60637 [295]

Sheng-Ping L. Hwang, Department of Pharmacological Sciences, Health Sciences Center, State University of New York, Stony Brook, NY 11794 [227]

Benita S. Katzenellenbogen, Department of Physiology and Biophysics, University of Illinois and College of Medicine, Urbana, IL 61801 [1, 201]

Kyoon E. Kim, Department of Physiology and Biophysics, University of Iowa, Iowa City, IA 52242 [159]

L.B. Kinsel, Department of Pathology, Duke University Medical Center, Durham, NC 27710 [279]

J.L. Kirkland, Division of Endocrinology, Department of Pediatrics, Baylor College of Medicine, Houston, TX 77030 [213]

Nancy L. Krett, Department of Medicine, University of Colorado, Health Sciences Center, Denver, CO 80262 [41]

John Kokontis, Ben May Institute, University of Chicago, Chicago, IL 60637 [53]

Deborah A. Lannigan, EHS Center, University of Rochester Medical School, Rochester, NY 14642 [187]

G. Leight, Department of Surgery, Duke University Medical Center, Durham, NC 27710 [279]

Shutsung Liao, Ben May Institute, Department of Biochemistry and Molecular Biology, University of Chicago, Chicago, IL 60637 [53]

T.H. Lin, Division of Endocrinology, Department of Pediatrics, Baylor College of Medicine, Houston, TX 77030 [213]

R.B. Lingham, Department of Pharmacology, University of Texas Medical School, Houston, TX 77225 [213]

Marc E. Lippman, Lombardi Cancer Research Center (S128), Georgetown University Medical Center, Washington, DC 20007 [243]

F. Logeat, Hormones et Reproduction, INSERM U. 135, Faculté de Médecine, Paris-Sud, 94275 Le Kremlin-Bicêtre, Cedex France [33]

D.S. Loose-Mitchell, Department of Pharmacology, University of Texas Medical School, Houston, TX 77225 **[213]**

H. Loosfelt, Hormones et Reproduction, INSERM U. 135, Faculté de Médecine, Paris-Sud, 94275 Le Kremlin-Bicêtre, Cedex France **[33]**

F. Lorenzo, Hormones et Reproduction, INSERM U. 135, Faculté de Médecine, Paris-Sud, 94275 Le Kremlin-Bicêtre, Cedex France **[33]**

K.S. McCarty, Jr., Department of Pathology, Duke University Medical Center, Durham, NC 27710 **[279]**

K.S. McCarty, Sr., Department of Biochemistry, Duke University Medical Center, Durham, NC 27710 **[279]**

Reid W. McNaught, Baylor College of Medicine, Houston, TX 77030 **[83]**

Richard A. Maurer, Department of Physiology and Biophysics, University of Iowa, Iowa City, IA 52242 **[159]**

Dirk B. Mendel, Department of Physiology, Dartmouth Medical School, Hanover, NH 03756 **[97]**

A. Migliaccio, Il Cattedra di Patologia Generale, I Facolta di Medicina e Chirurgia, Universita de Napoli, Italy **[133]**

E. Milgrom, Hormones et Reproduction, INSERM U. 135, Faculté de Médecine, Paris-Sud, 94275 Le Kremlin-Bicêtre, Cedex France **[33]**

M. Misrahi, Hormones et Reproduction, INSERM U. 135, Faculté de Médecine, Paris-Sud, 94275 Le Kremlin-Bicêtre, Cedex France **[33]**

V.R. Mukku, Department of Pharmacology, University of Texas Medical School, Houston, TX 77225 **[213]**

Allan Munck, Department of Physiology, Dartmouth Medical School, Hanover, NH 03756 **[97]**

Abhijit Nag, Baylor College of Medicine, Houston, TX 77030 **[83]**

Ann M. Nardulli, Department of Physiology and Biophysics, University of Illinois and College of Medicine, Urbana, IL 61801 **[201]**

Angelo C. Notides, Department of Biophysics, University of Rochester, Rochester, NY 14642 **[159,187]**

Sam Okret, Department of Medical Nutrition, Karolinska Institute, Huddinge University Hospital F69, S-141 86 Huddinge, Sweden **[65]**

C.A. Orengo, Department of Pharmacology, University of Texas Medical School, Houston, TX 77225 **[213]**

Eduardo Ortí, Department of Physiology, Dartmouth Medical School, Hanover, NH 03756 **[97]**

M. Pagano, Il Cattedra di Patologia Generale, Istituto di Patologia Generale, I Facolta de Medicina e Chirurgia, Universita de Napoli, Italy **[133]**

M. Perrot-Applanat, Hormones et Reproduction, INSERM U. 135, Faculté de Médecine, Paris-Sud, 94275 Le Kremlin-Bicêtre, Cedex France **[33]**

William B. Pratt, Department of Pharmacology, University of Michigan Medical School, Ann Arbor, MI 48109 **[119]**

Linnea D. Read, Department of Physiology and Biophysics, University of Illinois and College of Medicine, Urbana, IL 61801 **[201]**

Jeffrey L. Schwartz, Department of Radiation Oncology, University of Chicago, Chicago, IL 60637 **[295]**

Shirish Shenolikar, University of Texas Medical School, Houston, TX 77225 **[83]**

Philip L. Sheridan, Department of Pathology, University of Colorado, Health Sciences Center, Denver, CO 80262 **[41]**

Lynda I. Smith, Department of Physiology, Dartmouth Medical School, Hanover, NH 03756 **[97]**

Roy G. Smith, Baylor College of Medicine, Houston, TX 77030 **[83]**

G.M. Stancel, Department of Pharmacology, University of Texas Medical School, Houston, TX 77225 **[213]**

Per-Erik Strömstedt, Department of Medical Nutrition, Karolinska Institute, Huddinge University Hospital F69, S-141 86 Huddinge, Sweden **[65]**

Sarah Swift, Ben May Institute, University of Chicago, Chicago, IL 60637 **[53]**

Lisa L. Wei, Department of Medicine, University of Colorado, Health Sciences Center, Denver, CO 80262 **[41]**

Ann-Charlotte Wikström, Department of Medical Nutrition, Karolinska Institute, Huddinge University Hospital F69, S-141 86 Huddinge, Sweden **[65]**

David L. Williams, Department of Pharmacological Sciences, Health Sciences Center, State University of New York, Stony Brook, NY 11794 **[227]**

Alan P. Wolffe, Laboratory of Molecular Biology, NIDDK, NIH, Bethesda, MD 20892 **[171]**

Preface

Molecular Endocrinology and Steroid Hormone Action is the fourth in the International Cellular Endocrinology Symposium series held at the W. Alton Jones Cell Science Center in Lake Placid, New York. It is fitting that the meeting honors the contributions that Drs. Elwood Jensen and Jack Gorski have made to our knowledge of steroid hormone action. These two scientists have been at the forefront of research on steroid hormone action for more than 20 years. Indeed, the work of Drs. Jensen and Gorski has changed the way in which we think about steroid hormones. Through their research, and that of others, it has become apparent that steroid hormones act via specific receptors which translocate from the cytosol to the nucleus to change gene expression. This concept has become the central dogma of steroid hormone research.

The goal of the symposium was to highlight several areas of importance for steroid hormone action: steroid hormone receptor structure and function, steroid hormone receptors and their role as transacting factors, the posttranslational modification of receptors, the relationship between receptor activation and biological activity, and the use of steroid hormones in cancer. In achieving this goal, a diverse group of biologists, biochemists, molecular biologists, and clinicians was assembled. The diversity of the group served to underscore the extent to which the study of steroid hormone action has permeated many areas of biology and medicine. Perhaps more importantly, major unanswered questions were framed and refined.

It is our hope that this book reflects the level of scientific enthusiasm and discussion that was apparent during the meeting. A quick glance at the table of contents summarizes the organization of the text and will direct the reader to specific areas of interest. The section on structure and function provides an up-to-date comparison of estrogen, progesterone, glucocorticoid, and androgen receptors. In the following section, the role of steroid hormone receptors as transacting factors is compared to the current knowledge on transacting factors in the *Xenopus lavis* oocyte model to provide an overview of the current state of knowledge and directions for the future. The chapter on steroid hormone modification summarizes the current knowledge on an important and intriguing area of steroid hormone action, the role of receptor phosphorylation in hormonal regulation. The final two sections bring together knowledge on the role of steroid hormones in biology and pathobiology. The interplay of steroid hormones with growth factors is covered in several of the papers in these sections.

We hope that the reader will not only acquire a current knowledge of this exciting area of research, but will also gain a further appreciation for the contributions of Elwood Jensen and Jack Gorski as scientists and educators.

Gordon H. Sato
James L. Stevens

Acknowledgments

The Organizing Committee would like to thank the following corporations for their generous contribution to our Fourth International Symposium on Cellular Endocrinology.

Abbott Laboratories

Beckman Instruments, Inc.

Bristol-Myers Company

BRL, Life Technologies, Inc.

EI DuPont de Nemours & Co.

Eli Lilly Research Lab

Genentech, Inc.

Merck Sharp & Dohme Research Laboratories

Monsanto Company

Ortho Pharmaceutical Corp.

Pfizer Central Research

Pitman-Moore, Inc.

Sandoz Research Institute

Schering-Plough Corporation

Whittaker M.A. Bioproduct

We would also like to acknowledge the valuable skills and aid in the organization of this symposium from Carl A. Hamelin, Renee Johnston, and Front Office Staff. Also, thanks to Marina LaDuke, Polly Butler, and Carol Baine for their audio/visual assistance.

Due to their efforts our symposium was a success.

Symposium Participants

Jack Gorski. Elwood Jensen.

Alan Marks. Bob Dickson. Symposium participant. Eugene DeSombre.

Elwood Jensen. Pentti Siiteri. Herbert Jacobson. Symposium participant.

Shutsung Liao. Geoffrey Green. Jan-Åke Gustaffson.

Eugene DeSombre. Edwin Milgrom. Fernando Aurrichio.

Poster session.

Poster session.

Poster session.

Symposium participant. Jack Gorski. Symposium participant.

Elwood Jensen. Gordon Sato. James Stevens. Jack Gorski.

Molecular Endocrinology and Steroid
Hormone Action, pages 1–16
© 1990 Alan R. Liss, Inc.

A TRIBUTE TO JACK GORSKI: ESTROGEN RECEPTOR MODEL BUILDER
AND SCIENTIFIC FRIEND TO MANY

Benita S. Katzenellenbogen
Department of Physiology and Biophysics, University
of Illinois and University of Illinois College of
Medicine, Urbana, Illinois 61801

Jack Gorski has had a wonderful and enduring positive
influence on the many people he has trained. By his example
as an excellent scientist and clear thinker and through his
approachable, unassuming, and warm and friendly manner he
has set a fine example as a leader in the training of young
scientists. His graduate students, postdoctoral associates
and visiting scientists now number near 90 and many have
continued to populate the hormone-receptor field and have
themselves made important research contributions in
endocrinology and related fields. His associates and
students are represented as the many branches on the Gorski
tree (Figure 1), and many of these individuals are now
generating their own creative branches.

Something should be said also about the sound roots of
the Gorski tree (Fig. 1). Dr. Gorski received his
undergraduate education at the University of Wisconsin at
Madison. His explorations into the world of sex steroid
hormones and hormone action began during his M.S. and Ph.D.
studies at Washington State University at Pullman with Dr.
Robert Erb, where he characterized estrogens and progestins,
and their sources and activities during the reproductive
cycle and pregnancy in cattle. During this period, his
studies were enriched by his association as an NIH
Predoctoral Fellow with Dr. Leo Samuels and the Steroid
Biochemistry Training Program at the University of Utah.
His expanding interests in estrogen action were further
pursued during his return to the University of Wisconsin as
an NIH Postdoctoral Fellow with Dr. Gerald Mueller at the
McArdle Laboratory for Cancer Research. Here, his studies
with Dr. Mueller revealed the early effects of estrogens on
RNA and protein synthesis in the rat uterus.

POSTDOCTORAL ASSOCIATES: Donald Smith, G. Shyamala, Zygmunt Paszko, William Brennock, Ayalla Barnea, James Clark, George Giannopoulos, Thomas Ruh, Benita Katzenellenbogen, Gabriel Eilon, James Norris, George Stancel, Willian Beers, Tsuei-Chu Liu, Robin Leake, Jose Horacio Denari, Charles Frolik, Frank Gannon, Fred Stormshak, Richard Palmer, William Miller, Richard Maurer, Rita C. Manak, Mara Lieberman, Rich Ryan, Mark Richards, Conradin Meuli, Pierre Sartor, Leslie Hoffman, LeeAnn Baxter, Walter Hurley, James Weber, Linda Schuler, Wade Welshons, Carolyn Campen, Maria Recio, Mark Seyfred, Helga Ahrens, Frank Johnston, Madhubala Sonasekhar, Daniel Meier, Fern Murdoch.

Ph.D. STUDENTS: Jack Nicolette, William Noteboom, Angelo Notides, Jerry Reel, David Toft, Vytautas Uzgiris, Jack Barry, Patricia Coulson, Anthony DeAngelo, Paul Morris, Mary Sarff Forster, Peter Spooner, Neil Anderson, David Williams, Tom Spike, Kevin O'Connel, Jack Harris, Roger Stone, Richard Carlson, Margaret Shupnik, Julie Wiklund, Mary Slabaugh, Linda Vician, Gary Stack, Judy Kassis, Dennis Sakai, James Shull, Linda Durrin, Peter Rhode, Jeff Hansen, Laurie Shepel, Jane Walent, Kotaro Kaneko, Todd Winters, David Furlow, Mike Kladde, Mike Fritsch, Tamara Greco.

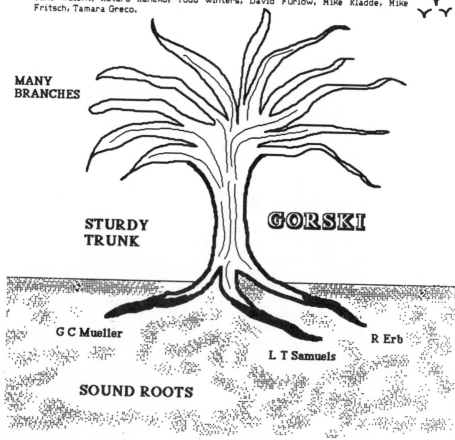

Figure 1. The Gorski tree, as of 1988.

In 1961, Jack Gorski assumed an Assistant Professorship position at the University of Illinois at Urbana-Champaign where he remained for 12 very productive years. During this period, Dr. Gorski attracted many excellent graduate students and postdoctoral fellows to his laboratory and made numerous observations of fundamental importance to our understanding of estrogen action (Figure 2). These include landmark studies on the identification of the receptor molecule for estrogen and the quantitative evaluation of its intracellular localization and interaction with nuclear binding sites; the characterization of the estrogen-induced synthesis of specific uterine proteins that documented, at an early time, the control of gene expression by estrogen; and, in an imaginative series of thoughtful studies, the analysis of the effects of estrogen on the synthesis of macromolecules in target cells. Dr. Gorski was recruited back to the University of Wisconsin-Madision in 1973 as a Professor in the Department of Biochemistry, Dairy Science and Meat and Animal Science, where he now holds a Paul H. Phillips Professorship and where his laboratory has continued to refine an evolving model of estrogen receptor action and to analyze growth regulation by estrogens and pituitary hormone syntheses and their modulation by estrogen.

Figure 3 presents a collage of just a sampling of some of these early, landmark papers on estrogen action. As shown in Figure 4, studies of Noteboom and Gorski in 1965 employing radiolabeled estradiol documented the nuclear localization of the hormone in the uterus after in vivo injection into rats, confirming the important independent observations of Elwood Jensen and Herbert Jacobson (1962). Studies by Toft and Gorski (1966) and Toft, Shyamala and Gorski (1967) (Figure 5) identified the hormone-binding receptor by sucrose gradient centrifugation techniques, and important studies by Williams and Gorski, Figure 6, presented an equilibrium model for hormone binding in the uterus documenting approximately 90% of the bound estradiol being firmly associated with the nucleus regardless of the degree of the fractional saturation of binding sites.

Concurrent studies in the early 1960s documented the stimulation of RNA polymerase activity (Figure 7) and a variety of early biosynthetic activities stimulated in the uterus by estradiol (Figure 8) and their dependence on protein synthesis. These findings implied that estrogens

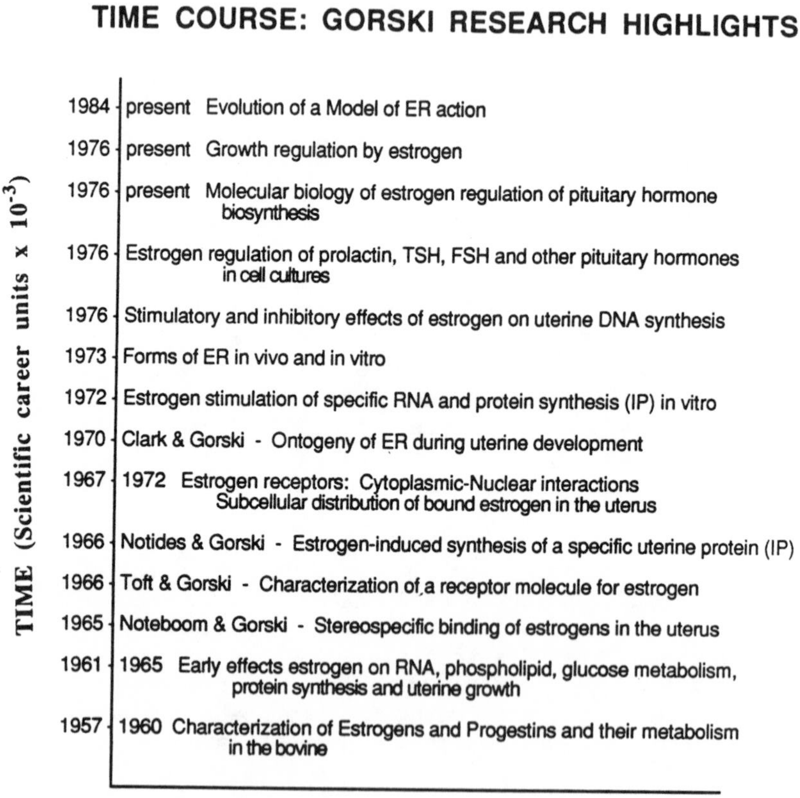

TIME COURSE: GORSKI RESEARCH HIGHLIGHTS

TIME (Scientific career units × 10^{-3})

1984 | present Evolution of a Model of ER action

1976 | present Growth regulation by estrogen

1976 | present Molecular biology of estrogen regulation of pituitary hormone biosynthesis

1976 | Estrogen regulation of prolactin, TSH, FSH and other pituitary hormones in cell cultures

1976 | Stimulatory and inhibitory effects of estrogen on uterine DNA synthesis

1973 | Forms of ER in vivo and in vitro

1972 | Estrogen stimulation of specific RNA and protein synthesis (IP) in vitro

1970 | Clark & Gorski - Ontogeny of ER during uterine development

1967 | 1972 Estrogen receptors: Cytoplasmic-Nuclear interactions Subcellular distribution of bound estrogen in the uterus

1966 | Notides & Gorski - Estrogen-induced synthesis of a specific uterine protein (IP)

1966 | Toft & Gorski - Characterization of a receptor molecule for estrogen

1965 | Noteboom & Gorski - Stereospecific binding of estrogens in the uterus

1961 | 1965 Early effects estrogen on RNA, phospholipid, glucose metabolism, protein synthesis and uterine growth

1957 | 1960 Characterization of Estrogens and Progestins and their metabolism in the bovine

Figure 2. Time course of Jack Gorski's research highlights.

ARCHIVES OF BIOCHEMISTRY AND BIOPHYSICS 111, 559-568 (1965)

Stereospecific Binding of Estrogens in the Rat Uterus

WILLIAM D. NOTEBOOM AND JACK GORSKI

Department of Physiology and Biophysics, University of Illinois, Urbana, Illinois

Reprinted from the Proceedings of the National Academy of Sciences
Vol. 55, No. 1, pp. 230-235. July, 1966.

ESTROGEN-INDUCED SYNTHESIS OF A SPECIFIC UTERINE PROTEIN*

By ANGELO NOTIDES† AND JACK GORSKI

DEPARTMENT OF PHYSIOLOGY AND BIOPHYSICS, UNIVERSITY OF ILLINOIS, URBANA

THE JOURNAL OF BIOLOGICAL CHEMISTRY
Vol. 247, No. 4, Issue of February 25, pp. 1299-1305, 1972
Printed in U.S.A.

Estrogen Action *in Vitro*

INDUCTION OF THE SYNTHESIS OF A SPECIFIC UTERINE PROTEIN*

BENITA S. KATZENELLENBOGEN AND JACK GORSKI

From the Departments of Physiology and Biophysics, and Biochemistry, University of Illinois, Urbana, Illinois 61801

Reprinted from the Proceedings of the National Academy of Sciences
Vol. 58, No. 6, pp. 1574-1581. June, 1968.

A RECEPTOR MOLECULE FOR ESTROGENS: ISOLATION FROM THE RAT UTERUS AND PRELIMINARY CHARACTERIZATION*

By DAVID TOFT† AND JACK GORSKI

DEPARTMENT OF PHYSIOLOGY AND BIOPHYSICS, UNIVERSITY OF ILLINOIS, URBANA

Excerpta Medica International Congress Series No. 219
HORMONAL STEROIDS
Proceedings of the Third International Congress, Hamburg, September 1970

CONTROL OF GENE EXPRESSION: AN EARLY RESPONSE TO ESTROGEN*

J. GORSKI, A. B. DeANGELO** and A. BARNEA***

GYNECOLOGIC ONCOLOGY 2, 249-258 (1974)

Estrogen Action in the Uterus: The Requisite for Sustained Estrogen Binding in the Nucleus[1]

JACK GORSKI AND BETTY RAKER

Departments of Biochemistry and Meat and Animal Science, University of Wisconsin, Madison, Wisconsin 53706

Proc. Nat. Acad. Sci. USA
Vol. 69, No. 11, pp. 3464-3468, November 1972

Kinetic and Equilibrium Analysis of Estradiol in Uterus: A Model of Binding-Site Distribution in Uterine Cells

DAVID WILLIAMS* AND JACK GORSKI

Departments of Physiology and Biophysics and Biochemistry, University of Illinois, Urbana, Ill. 61801

Figure 3. Some important early publications from the Gorski Lab.

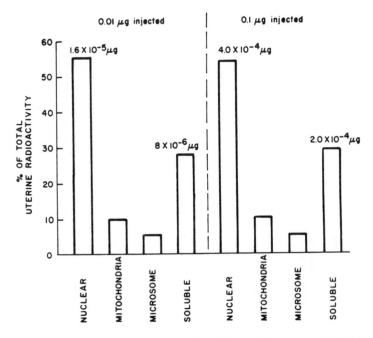

Figure 4. The distribution of radioactive estradiol in subcellular fractions of rat uteri after in vivo injection of estradiol. (From Noteboom and Gorski, 1965)

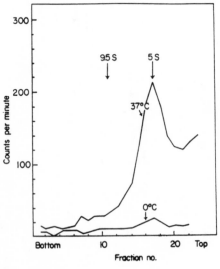

Figure 5. Sucrose gradient patterns of 0.3M KCl extracts of nuclei after whole uterine tissue incubations with ^{3}H-estradiol. (From Toft et al., 1967)

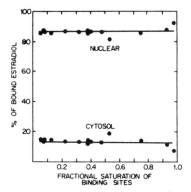

Figure 6. Equilibrium distribution of filled nuclear and cytosol binding sites. (From Williams and Gorski, 1972)

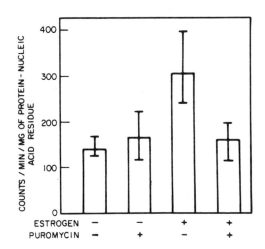

Figure 7. Effect of estradiol and puromycin on the activity of uterine RNA polymerase. (From Noteboom and Gorski, 1963)

might stimulate the synthesis of specific proteins, and
these ideas led to the experiments by Notides and Gorski
(1966) that documented the estrogen-induced synthesis of a
specific uterine protein early after estrogen exposure
termed "induced protein" or IP. Subsequent studies
demonstrated this induced protein synthesis to be dependent
on RNA and protein synthesis and showed it to be a rapid and
primary response to the hormone (Barnea and Gorski, 1970;
Deangelo and Gorski, 1970). The observation that IP
synthesis could be stimulated by low, physiological
concentrations of estrogens in intact uteri in vitro (Figure
9), then allowed a careful study of the quantitative
relationships (Figure 10) between IP synthesis and nuclear
receptor occupancy for a variety of different estrogens
(Katzenellenbogen and Gorski, 1972; Ruh et al., 1973;
Katzenellenbogen and Gorski, 1975). Figure 11 shows a
sampling of some of the important studies from the Gorski
lab in the mid 1970s since it indicates the start of what
has become a continuing interest in estrogen action both at
the pituitary and uterine levels.

In all of these studies, Jack Gorski has maintained a
good sense of what is biologically important and of how to
approach these questions experimentally. He has brought
modern techniques of molecular biology to bear on important
physiological problems in endocrinology, but he has never
let techniques rule his science to the exclusion of
addressing difficult, central physiological issues.

Perhaps Jack Gorski's greatest scientific contribution
has been in the refinement and continuing evolution of a
model for estrogen receptor action. Here, he has
continually modeled, revised, and remodeled the estrogen
receptor model as new data dictate. A time-course evolution
of the estrogen receptor model is presented schematically in
Figure 12. The early model in Figure 13, indicating the
intimate association of the estrogen receptor with chromatin
to activate RNA transcription and the sustained output model
of estrogen action, whereby continuing action of receptors
in the nucleus is required for long-term responses to
estrogen, remain very important aspects. The enucleation
studies of Welshons et al. in 1984 forced a reevaluation of
the intracellular localization of the unoccupied estrogen
receptor and a realization (Figure 14), documented also by
others using immunocytochemistry (King and Greene, 1984),
that the unliganded receptor is largely nuclear, although

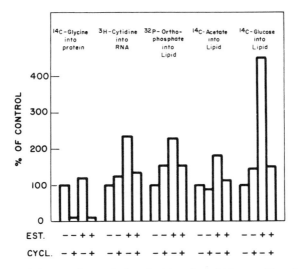

Figure 8. Effect of cycloheximide (cycl.) on the 2-hour uterine response to estradiol. (From Gorski et al., 1965)

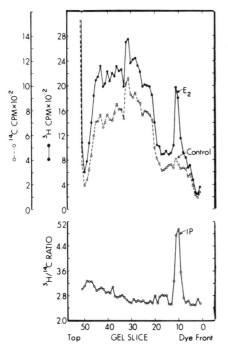

Figure 9. Electrophoretic distribution on gels of uterine soluble proteins including the estrogen induced protein, IP, synthesized in vitro following a 1-hour in vitro incubation of uteri with 3.7×10^{-8}M estradiol. (From Katzenellenbogen and Gorski, 1972)

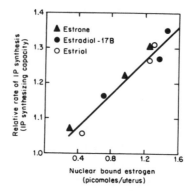

Figure 10. Relative rate of estrogen induced protein (IP) synthesis as a function of nuclear bound estrogen. (From Katzenellenbogen and Gorski, 1975)

Proc. Natl. Acad. Sci. USA
Vol. 75, No. 12, pp. 5946–5949, December 1978
Biochemistry

Estrogen control of prolactin synthesis *in vitro*

(primary pituitary cell culture/immunoprecipitation)

MARA E. LIEBERMAN*, RICHARD A. MAURER†, AND JACK GORSKI*

* Departments of Biochemistry and Animal Science, University of Wisconsin-Madison, Madison, Wisconsin 53706; and † Department of Physiology and Biophysics, University of Iowa, Iowa City, Iowa 52242

Communicated by Elwood V. Jensen, September 25, 1978

Evidence for a Discontinuous Requirement for Estrogen in Stimulation of Deoxyribonucleic Acid Synthesis in the Immature Rat Uterus*

(*Endocrinology* 103: 240, 1978)

JACK HARRIS† AND JACK GORSKI

Departments of Biochemistry and Animal Science, University of Wisconsin, Madison, Wisconsin 53706

Stimulatory and Inhibitory Effects of Estrogen on Uterine DNA Synthesis[1,2]

(*Endocrinology* 99: 1501, 1976)

FREDRICK STORMSHAK,[3] ROBIN LEAKE,[4] NANCY WERTZ, AND JACK GORSKI[5]

Departments of Biochemistry and Animal Science, University of Wisconsin, Madison 53706

Figure 11. Some important publications from the Gorski lab in the mid-1970s.

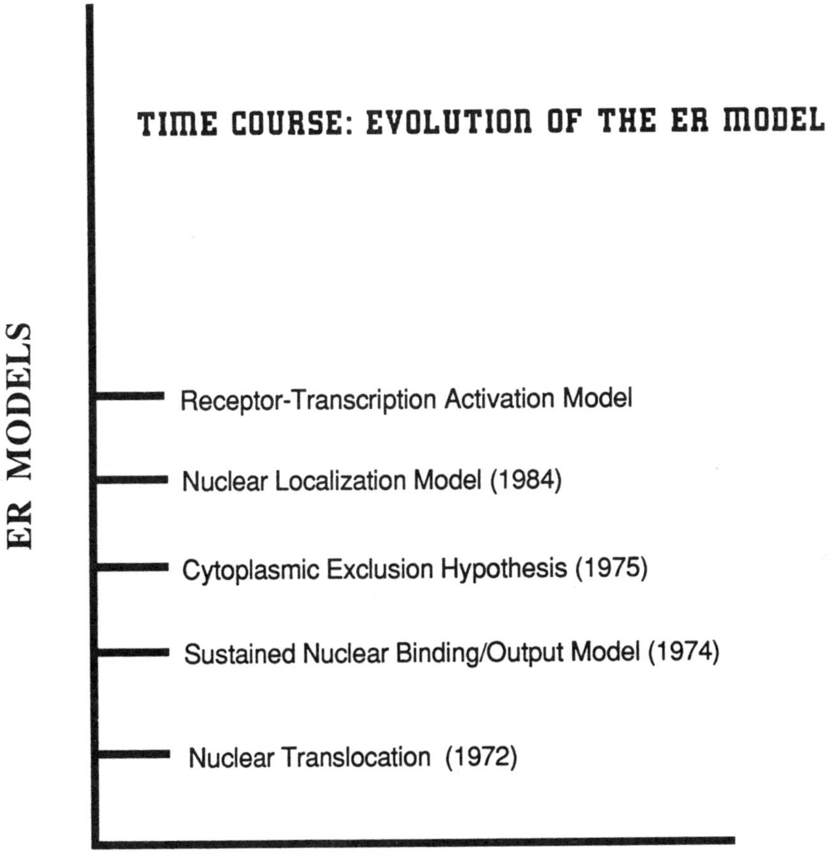

TIME COURSE: EVOLUTION OF THE ER MODEL

ER MODELS

Receptor-Transcription Activation Model

Nuclear Localization Model (1984)

Cytoplasmic Exclusion Hypothesis (1975)

Sustained Nuclear Binding/Output Model (1974)

Nuclear Translocation (1972)

Figure 12. Time course evolution of the estrogen receptor model.

more loosely associated, before estrogen interaction with receptors. Continuing studies of Gorski of his associates have documented a variety of forms of the estrogen receptor in cells (Figure 14), and studies on regulation of the prolactin gene by his laboratory have resulted in even newer, transcription activation models in which the estrogen receptor is required for generation of an active transcription complex, although the details of this are still to be more thoroughly explored (Figure 14).

Everyone who knows Dr. Gorski finds him to be an especially thoughtful, honest and warm individual, always approachable, with an exceptional concern for his students as people and as scientists. He exudes an intense, yet almost boyish enthusiasm and interest in science that is highly infectious to those who work with him. His identification and characterization of estrogen receptors and studies on their mechanism of action have served as a model for understanding the actions of other steroid hormones in a variety of other systems. In addition, his broad base in biology and breadth of interest in endocrinology, biochemistry and physiology have had an enduring effect on the scientific approach and careers of the many scientists who have worked with him.

Dr. Gorski's administrative and teaching activities also reflect his interest and concern with the development of young scientists and the future of science in this country. He has served as chairman of the Membership Committee of the Endocrine Society and the American Society of Biological Chemists and he presently serves on the American Cancer Society Committee on Personnel for Research and on the Committee on a National Strategy for Biotechnology in Agriculture of the National Research Council. He has also served on numerous review panels of the National Institutes of Health and the American Cancer Society and has been on the editorial boards of many journals, and he is widely sought after because of his broad perspective in approaching and appreciating science and because of his fair-minded spirit. Throughout all of this, he has also influenced many undergraduate and graduate students at the University of Illinois and University of Wisconsin through exposure to his very popular endocrinology and cell regulation courses.

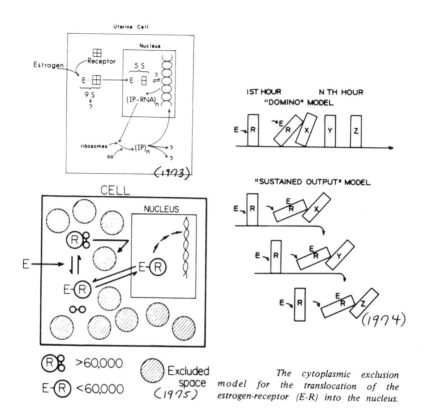

The cytoplasmic exclusion model for the translocation of the estrogen-receptor (E-R) into the nucleus.

Figure 13. Some early J. Gorski estrogen receptor models.

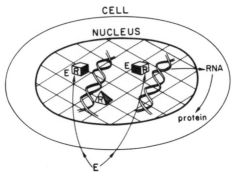

"New" model of estrogen receptor. R, Receptor; E, estrogen.

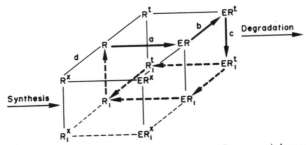

Model of potential receptor forms. E, Estrogen; R, unoccupied receptor; ER, nontransformed/occupied receptor; R^t, transformed receptor; R^k, modified receptor.

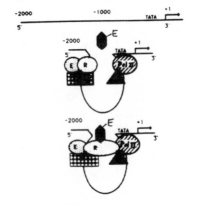

Figure 14. Some more recent J. Gorski estrogen receptor models.

Dr. Gorski received the Ernst Oppenheimer Memorial Award of the Endocrine Society in 1971 for early recognition of his outstanding work on the mechanisms of estrogen action, and he was honored in 1968 and again in 1985 to present the work of his laboratory at the Laurentian Hormone Conference. In 1986, he received the Gregory Pincus Medal and Award and was elected to the American Academy of Arts and Sciences. Based on his distinguished leadership in research, teaching and in the training of scientists, the Endocrine Society honored Jack Gorski with its Robert H. Williams Distinguished Leadership Award in 1987.

It is very befitting that both Elwood Jensen and Jack Gorski are being honored for their contributions in the identification of estrogen receptors and elucidation of their important roles in target cells. It is important to note also that the friendship between Jack and Elwood and their openness to new people entering the estrogen receptor field was very important in making it an area in which young people wanted to work and felt welcomed and encouraged. This was indeed important in the development of the many scientists they have trained.

Everyone who has had the opportunity to work with Jack Gorski feels very privileged and a part of the "Gorski clan". Dr. Gorski has clearly been a leader in the training and development of endocrinologists, many of whom are themselves now quite eminent and influential in this field. Through his insightful questions, broad interest in biology and his warm and unpretentious personality, he has nurtured the development of his many associates into productive and inquisitive scientists. On behalf of your many former associates, we salute this wonderful scientist and person and wish him all the best in the future. We have no doubt that the Gorski tree will continue to sprout additional productive branches.

REFERENCES

Barnea A, Gorski J (1970). Estrogen induced protein: Time course of synthesis in vivo. Biochemistry 9:1899-1904.
DeAngelo AB, Gorski J (1970). The role of RNA synthesis in the estrogen induction of a specific uterine protein. Proc Natl Acad Sci USA 66:693-700.

Gorski J, Noteboom WD, Nicolette J (1965). Estrogen control of the synthesis of RNA and protein in the uterus. J Cell Comp Physiol 66, Suppl. 1:91-109.

Jensen EV, Jacobson HI (1962). Basic guides to the mechanism of estrogen action. Recent Prog Hormone Res 18:387-414.

Katzenellenbogen BS, Gorski J (1972). Estrogen action in vitro: Induction of the synthesis of a specific uterine protein. J Biol Chem 247:1299-1305.

Katzenellenbogen BS, Gorski J (1975). Estrogen actions on syntheses of macromolecules in target cells. In: Biochemical Actions of Hormones, Vol III. G Litwack, ed, Academic Press, New York, pp 187-243.

King WJ, Greene GL (1984). Monoclonal antibodies localize estrogen receptor in the nuclei of target cells. Nature 307:745-747.

Noteboom W, Gorski J (1963). An early effect of estrogen on protein synthesis. Proc Natl Acad Sci USA 50:250-255.

Noteboom W, Gorski J (1965). Stereospecific binding of estrogens in the rat uterus. Arch Biochem Biophys 111:559-568.

Notides A, Gorski J (1966). Estrogen-induced synthesis of a specific uterine protein. Proc Natl Acad Sci 56:230-235.

Ruh TS, Katzenellenbogen BS, Katzenellenbogen JA, Gorski J (1973). Estrone interaction with the rat uterus: In vitro response and nuclear uptake. Endocrinology 92:125-134.

Toft D, Gorski J (1966). A receptor molecule for estrogen: Isolation from the rat uterus and preliminary characterization. Proc Natl Acad Sci USA 55:1574-1580.

Toft D, Shyamala G, Gorski J (1967). A receptor molecule for estrogens: Studies using a cell-free system. Proc Natl Acad Sci USA 57:1740-1743.

Welshons W, Lieberman ME, Gorski J (1984). Nuclear localization of unoccupied estrogen receptors: Cytochalasin enucleation of GH_3 cells. Nature 307:747-749.

Williams D, Gorski J (1972). Kinetic and equilibrium analysis of estradiol binding in the uterus: A model of binding site distribution in the uterine cells. Proc Natl Acad Sci USA 69:3464-3468.

Molecular Endocrinology and Steroid
Hormone Action, pages 17–29
© 1990 Alan R. Liss, Inc.

ESTROGENS, RECEPTORS AND CANCER: THE SCIENTIFIC CONTRIBUTIONS
OF ELWOOD JENSEN.

Eugene R. DeSombre

Ben May Institute, University of Chicago
Chicago, IL 60637

INTRODUCTION

This chapter will document the contributions of Elwood
Jensen to knowledge of steroid hormone action in normal and
neoplastic tissues. Although Elwood Jensen would be the
first to acknowledge the important contributions of many other scientists to the advances which are detailed here, because of space limitations I have tried to highlight Elwood's
distinctive contributions without always placing them in the
context of other developments which may have influenced his
thinking or experimental approach. Those contributions of
others can be better appreciated from the more extensive reviews of the field that Elwood has written (Jensen and Jacobson, 1962; Jensen and DeSombre, 1973; Jensen et al., 1982).
There is no doubt that Elwood Jensen laid the groundwork for
the modern era of steroid hormone action, he contributed numerous insightful and seminal discoveries to our understanding of this field and was responsible for the application of
this knowledge to aid clinicians in treating breast cancer
patients.

EARLY RESEARCH

It has been said that success comes to those who are
prepared. It is interesting to see how training, environment
and experience prepared Elwood Jensen to ask the questions
which changed the direction of research on steroid hormones.
After undergraduate education at Wittenberg College in Ohio,
Elwood came to the University of Chicago where he carried out
his doctoral research with the distinguished organic chemist,
M.S. Kharasch. Following world war II, with support of a
Guggenheim fellowship, he studied synthetic steroid chemistry
in Zurich. On his return he was attracted to the laboratory
of Dr. Charles Huggins, the renowned surgeon who would subsequently receive the Nobel Prize for his contributions to the
control of hormone-dependent cancers. Dr. Huggins, an advo-

cate of multidisciplinary approaches to medical research, had gathered a diverse group of scientists at the University of Chicago to study basic biology related to cancer. In his laboratory, Elwood studied proteins, sulfhydryl groups and cancer, themes which were to return in later research.

The early work of Huggins on prostatic cancer had gained the attention of a Mobile businessman, Ben May, whose support led to the establishment in 1951 of a new department, the Ben May Laboratory for Cancer Research. Here Elwood found an ideal environment to continue his basic and clinical research interests. Armed with his postdoctoral experience in steroid chemistry, he and Huggins began to study estrogen action, which would become his lifelong interest. They reported on the relation of structure to function for a number of estrogenic steroids (Huggins et al., 1954; Huggins and Jensen, 1954) and assessed the biologic activity of certain polyoxygenated steroids, such as estriol, for which they coined the term "impeded estrogens," since these agents showed depressed maximum biologic activity compared to that of estradiol, and, on coadministration, reduced the full response of active estrogens (Huggins and Jensen, 1955). These substances were thus among the first to suppress, at least partially, biologic responses to estrogenic hormones without inhibiting general biosynthetic pathways.

During the early 1950's, Elwood attracted a group of outstanding students and research associates in organic chemistry, and developed a number of novel chemical reactions for the synthesis of phosphonate analogs of biologically important phosphates, and of fluorinated derivatives of steroid hormones to determine how the chemical modifications affected biologic activity. But these investigations were all preliminary to his major contribution, the discovery of steroid hormone receptors.

STUDIES OF ESTROGEN ACTION

From his perspective as a chemist, Elwood recognized that to understand how such minute amounts of estrogenic hormones can elicit their spectacular growth response in tissues of the reproductive tract one had to know what the actual "reactants" were. Though others had attempted to use ^{14}C-labeled steroids to follow the fate of administered hormone in experimental animals, such probes did not have the sensitivity required to detect physiologic amounts of estrogen in the target tissues. In collaboration with Herbert Jacobson and Gopi Gupta, he set out to synthesize radiolabeled estradiol of high specific activity, a goal accomplished by the catalytic reduction of 6,7-dehydroestradiol with carrier-free tritium (Jensen and Jacobson, 1960). It was also necessary to develop a new procedure to measure tritium in animal tissues (Jacobson et al., 1960). Using these tools, in late 1957 they initiated a now classical series of experiments to determine what actually happens to the estradiol molecule as it initiates growth in the immature rat uterus.

At the International Congress of Biochemistry in Vienna in 1958 Elwood reported the first results of their studies on the incorporation of tritium-labeled estrogens in animals. As often happens important events that change the direction of science may go largely unnoticed at the time. At this congress, his presentation came at the same time as a major symposium on hormone action, which included discussion of the then popular view of how estrogens act, involving the reversible oxidation-reduction of estradiol and the transhydrogenase-linked, phosphopyridine nucleotide cofactor-dependent energy transfer. While only a handful of the attendees were present to hear Elwood's paper, it heralded a new approach to molecular endocrinology and was clearly the first building block of the present edifice of knowledge about hormone action, receptor biochemistry and gene regulation in normal and neoplastic tissues.

Illustrated in Figure 1 is what has become a classic experiment which provided the first evidence for specific steroid-binding components in target tissues for the estrogens. Elwood reported that, after administration of physiologic amounts of tritiated estradiol to immature rats, the uterus, vagina and anterior pituitary showed a continued, substantial uptake and prolonged retention of radioactivity, consistent with the presence in such tissues of a specific binding component for estrogen (Jensen and Jacobson, 1960). Other tissues, despite a sometimes initially high level of radioactivity, rapidly lost the tritium labeled ligands in a manner

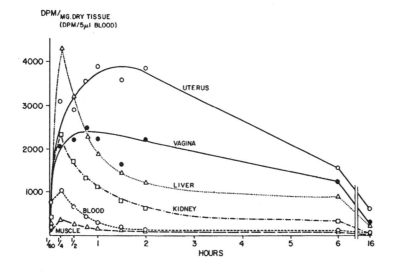

Fig 1. Concentration of radioactivity in immature rat tissues after subcutaneous injection of 11.5 μCi (98 ng) of 6,7 ^{3}H estradiol in saline. From Jensen and Jacobson (1960).

paralleling that of the blood. While the blood and non-target tissues contained a mixture of metabolites, Elwood showed that in the target tissues the radioactivity was largely unchanged estradiol (Jensen et al., 1966). To rule out the possibility of reversible oxidation-reduction between estradiol and estrone, with the equilibrium amount of estrone being too small to detect, he synthesized 17α-tritiated estradiol and showed that, when administered to rats, the estradiol bound in the uterus did not lose its tritium. Thus it was established that estradiol elicits growth of the immature rat uterus without itself undergoing metabolic change. (Jensen and Jacobson, 1962), thereby putting to final rest the transhydrogenation hypothesis for estrogen action.

At this time there was no direct evidence that the association of the steroid with the binding substances played a role in biologic response, so that one could refer to these as "receptors". Elwood coined the term "estrophilin" to describe the estrogen-binding components of target tissues (Jensen, 1967). He later showed that estrophilin was a true receptor by using varying doses of the antiestrogen, nafoxidine, to establish a correlation between the inhibition of binding and the inhibition of growth in the rat uterus (Jensen, 1965; Jensen et al., 1972). Using increasing doses of labeled estradiol, he demonstrated that the specific uptake and retention of hormone by the target tissues was a saturable phenomenon (Jensen et al., 1967a), and his in vitro studies showed that the interaction of estradiol with this putative receptor was reversible and sensitive to sulfhydryl reagents (Jensen et al., 1967c).

Based on the relationship between "cytosolic" and "nuclear" binding, the depletion of cytosolic receptor as estrogen is concentrated in the uterine nuclei in vivo, the observation that cytosolic receptor is required for the formation of the nuclear receptor in isolated nuclei, and the temperature-dependent redistribution of estrogen binding in uteri previously exposed to estradiol in the cold, Jensen and co-workers proposed a two-step interaction pathway in which the tightly bound nuclear hormone-receptor complex is derived from an initially formed cytosolic complex (Jensen et al., 1968). About the same time, Gorski and coworkers independently proposed a similar mechanism in which the nuclear receptor was produced from the cytosolic receptor (Gorski et al., 1968). By using the valuable technique of sedimentation analysis, first introduced by Toft and Gorski (1966) for the characterization of receptor complexes, Elwood showed that the estradiol-receptor complex extracted from uterine nuclei and be distinguished from that present in the cytosol by centrifugation in a salt-containing sucrose gradient (Figure 2), where the nuclear complex sediments at about 5S and the cytosol complex at 4S (Jensen et al., 1969; Jensen and DeSombre, 1973). His laboratory demonstrated the hormone-induced, temperature-dependent conversion of the 4S estradiol-receptor complex to the 5S form in vitro (Jensen et al., 1971b), and showed that only the 5S complex had the ability to bind

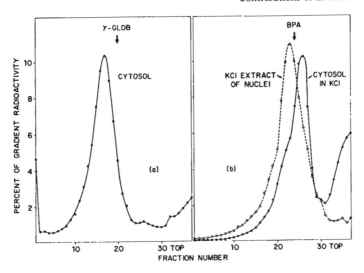

Fig 2. Sedimentation patterns of ^{3}H E2 receptor complexes of rat uterine cytosol and nuclear extract from uteri of immature rats excised 1 hr after the subcutaneous injection of ^{3}H E2 in vivo. Cytosol was made 5 nM with additional ^{3}H E2. Gradients are: left, 10-30% sucrose, low salt; right, 5-20% sucrose with 400 mM KCl (from Jensen and DeSombre, 1973).

tightly to isolated nuclei (DeSombre et al., 1975). With Suresh Mohla he showed that only the 5S complex had the ability to increase the RNA-synthesizing capacity of nuclei from target cells but not non-target cells (Mohla et al., 1972). Thus the hormone-induced change of the 4S "native" receptor to the 5S nuclear form, originally known as receptor "transformation", now usually called receptor "activation", was recognized as a key function of the steroid hormone responsible for the conversion of the intracellular receptor protein to a biochemically functional regulator of gene expression (Jensen and DeSombre, 1973; DeSombre et al., 1975).

As results of studies on the interaction of other steroid hormones with their target tissues appeared, Elwood recognized many similar features and suggested a common mechanism for steroid hormones action mediated by the specific intracellular receptors, different from the mechanism for peptide hormones acting through membrane receptors (Jensen et al., 1971b). This early insight brought unification to the field although it would be 15 years before the similar structural domains and the highly conserved DNA binding regions of the various steroid receptors would be recognized as the basis of the common mechanisms of action of the steroid hormones. Thus while Elwood focused his studies on estrogen action, his groundbreaking experiments and insights had a significant influence on progress in the entire field of

```
┌─────────────────────────────────────────────────────────────────────┐
│                         Major Contributions                          │
│                                                                       │
│  1. Specific uptake and retention of estrogens by target tissues      │
│     in vivo                                                           │
│  2. Estradiol acts without metabolic conversion                       │
│  3. Retention of estrogens is a saturable phenomenon (receptors)      │
│  4. Intact sulfhydryl groups needed for estrogen binding              │
│  5. Autoradiography for subcellular localization of 3H estradiol      │
│  6. Two-step interaction mechanism                                    │
│  7. Activation of ER needed for tight nuclear binding                 │
│  8. Temperature-dependent activation of ER in vitro                   │
│  9. 4S to 5S conversion of ER relates to tight nuclear binding        │
│  10. ER stimulation of RNA polymerase activity of uterus              │
│                                                                       │
│                             Coworkers                                 │
│  P.I. Brecher, V. Colucci, E.R. DeSombre, J.W. Flesher, G.N. Gupta,   │
│  M. Ikeda  A. Hughes, D.J.Hurst, H.I. Jacobson, P.W. Jungblut,        │
│  T. Kawashima  S. Mohla, H.G. Neuman, N.N. Saha, D. Shiplacoff,       │
│  S. Smith, W.E. Stumpf  T. Suzuki                                     │
└─────────────────────────────────────────────────────────────────────┘
```

Fig. 3. Contributions and coworkers of E.V. Jensen in studies on the mechanism of estrogen action

steroid hormone action. Figure 3 summarizes many of the contributions of Elwood Jensen's laboratory to studies of the mechanism of estrogen action, and lists most of the individuals who collaborated with him in these studies.

PURIFICATION OF ESTROGEN RECEPTORS; ANTIBODY PRODUCTION

Another major emphasis of Jensen's laboratory was the purification of the estrogen receptor (ER). These studies were long and arduous, lasting for more than a decade, but eventually successful (Gorell et al., 1977), overcoming the problems of the minute concentrations of receptor in target tissues and the lability of the receptor protein in tissue extracts. The contributions of his laboratory, and the coworkers involved in these efforts are shown in Fig 4. These accomplishments include the first attempt to use a steroid affinity column for the purification of a steroid receptor (with Peter Jungblut - Jungblut et al., 1967) and the first completely successful method for purification of estrogen receptor by affinity chromatography (Greene et al., 1980). With Geoffrey Greene, recognition that antibodies to estrophilin form soluble immune complexes led to the preparation of the first definitive antibodies to a steroid hormone receptor (Greene et al., 1977) and later to a library of monoclonal antibodies to estrogen receptors from human breast cancer cells (Greene et al., 1980, 1984). These substances not only permitted the purification of estrophilin by antibody affinity columns (Greene, 1984), but they provided reagents that were crucial in the development of immunoassay procedures for the measurement of estrogen receptors in breast cancers (discussed below), the immunochemical lo-

---

Major Contributions

Isolation and purification of "cytosolic" ER of calf uterus
Isolation and purification of "nuclear" ER of calf uterus
Preparation of first well-characterized Ab to a steroid receptor
Isolation and purification of ER of MCF-7 breast cancer cells
Preparation of first monoclonal AB to a steroid receptor
Definition of functional domains of estrogen receptor

Coworkers

J.P. Chabaud, L.E. Closs, E.R. DeSombre, T.A. Gorell, F.W. Fitch, H. Fleming,
G.L. Greene, P.W. Jungblut, C. Nolan, G.A. Puca, N. Sobel, S. Tanaka

---

Fig 4. Contributions and coworkers of E.V. Jensen on the purification of estrogen receptors and preparation of antibodies to ER

calization of estrogen receptor in nuclei of untreated cells (King and Green 1984), as well as for the cloning of the cDNA for estrophilin (Walter et al., 1985; Greene et al., 1986) and for definition of the functional domains in the estrogen receptor molecule (Greene et al., 1984, 1986).

ESTROGEN RECEPTORS AND BREAST CANCER

It was quite natural that in the Ben May Laboratory for Cancer Research Elwood would try to apply the new knowledge he had gained on the mechanism of action of estrogens in normal tissues to cancer. Huggins had shown that some, but not all, patients obtained dramatic remission of advanced breast cancer following adrenalectomy (Huggins and Bergenstal, 1952). However, there was no method to identify which patients were likely to respond. Elwood turned to the rat model system that Huggins had developed for breast cancer and found that the ovarian responsive rat mammary cancers resembled normal estrogen target tissues in their ability to concentrate estrogens in vivo and in vitro (Jungblut et al., 1967). With the aid of Drs Block and Ferguson, Elwood obtained some human breast cancers to study in vitro and identified 2 types of breast cancers, those with and those without antiestrogen-inhibitable uptake of tritiated estradiol (Jensen, et al., 1967b; Jensen et al., 1971a). Furthermore, it appeared that only those patients whose cancers showed inhibitable uptake responded to endocrine ablative therapy (Jensen et al., 1971a). While this assay required substantial amounts (1 g) of fresh tissue, limiting its application, it provided the first evidence that assay of the ER content of human breast cancer could be useful to identify the group of patients who were most likely to benefit from endocrine therapy. Subsequently, he applied sedimentation analyses of ER in low salt extracts of breast cancer homogenates which circumvented the major limitations of the original assays so that most breast cancers could be assayed for ER content (Jensen et al., 1971a, 1976).

After the original report of the correlation of the

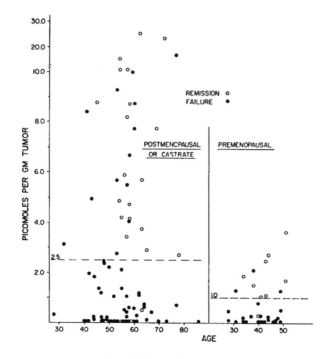

Fig. 5 Correlation of tumor cytosol ER content with response to endocrine therapy for 133 patients with advanced breast cancer. ER was determined by sedimentation analysis of tumor cytosol using 0.5 nM tritiated estradiol, and results were corrected to total binding capacity as described elsewhere (Jensen et al., 1976).

presence of ER in breast cancers with patient response to endocrine therapy, laboratories throughout the world began assaying breast cancers for ER. The results from the worldwide trials confirmed the usefulness of the ER assays for the prediction of response to endocrine therapy (McGuire et al., 1975; DeSombre et al., 1979) and led to the recognition of ER as a prognostic factor in breast cancer (Knight et al., 1977). While almost all the patients who responded to endocrine therapy had ER+ cancers, not all patients with ER+ cancers responded to such therapy. Therefore Elwood reevaluated the data from his clinical studies and discovered that those patients with low levels of ER in their cancers, like those lacking ER, were unlikely to respond and furthermore, the quantity of ER related to response rate, Figure 5 (Jensen et al., 1976; DeSombre and Jensen, 1980). Hence consideration of the quantity of ER was important for assessing the likelihood of response to endocrine therapy.

With the wider application of ER determinations in breast cancers and the need for quantitative measurements, titration assay methods using Scatchard plots became stan-

---

Major Contributions

Ovariectomy responsive rat mammary cancers have ER
Specific uptake of estrogens by slices of breast cancer in vitro
Sedimentation analysis of human breast cancer extracts to assay ER
Correlation of ER presence with breast cancer response to endocrine Rx
Quantity of ER relates to response to endocrine therapy
Antibodies to ER for assay of estrogen receptors
    Quantitative ER Assay with antibodies (Enzyme immunoassay, ER-EIA)
    Immunocytochemical ER assay (ER-ICA)

Coworkers

G.E. Block, E..R. DeSombre, R.S. Ellis, D. Ferguson, G.L. Greene, P.W. Jungblut
W.J. King, K.A. Kyser, S. Smith

Fig. 6. **Contributions and coworkers on ER in breast cancer**

dard. These new assays also depended upon the binding of estrogen to the receptor and were often problematic to reproduce from one laboratory to another. However, with the production of antibodies to ER, it became possible for the first time to develop methods for measurement of ER not dependent upon the binding of steroid. In collaboration with Abbott Laboratories, an enzyme immunoassay, based on the monoclonal antibodies to ER, was developed and has been certified by the FDA for the assay of ER in breast cancer extracts. Early indications suggest that there is better reproducibility and quality control for this immunoassay for ER and that it shows excellent correlation with conventional steroid binding assays for ER in breast cancers (Leclerq et al., 1986; Jordan et al., 1986).

It was also desirable to be able to assess the heterogeneity of ER expression within the breast cancer tissue to recognize the presence of any sub-population of receptor-negative cancer cells in a predominantly receptor-positive tumor. Such assessment was obviously not possible with methods which depend upon homogenization of the tissue. But with immunocytochemical assay methods, referred to earlier, one could assess such non-uniform distribution in cancers and equally important, determine whether the ER present was actually expressed in the cancer cells themselves. The initial report from Elwood's laboratory on the use of ER immunocytochemistry (ER-ICA) in breast cancer tissue, showed an excellent correlation between ER-ICA and conventional ER assays, and also suggested its potential for assessing ER heterogeneity (King et al., 1985).

The contributions of Elwood's laboratory in the area of breast cancer are listed in Figure 6. The establishment of the clinical usefulness of ER assays in breast cancer not only helped clinicians treat patients with this disease, but also established an important precident for application of emerging basic knwledge to the clinical setting. His early studies were only possible because of the support and dedication of several key surgeons who insured that tissue and patient response information were available. The success of these studies and the clinical usefulness resulting from the application of these basic laboratory advances to clinical medicine broke ground for subsequent developments in a variety of areas, so that now clinical applications derived from emerging knowledge in molecular and cellular biology are sought out and encouraged in the recognition that there may be a real benefit for patient care.

EPILOGUE

It would not be appropriate to end this discussion of Elwood Jensen's contribution to science without a few words about Elwood himself. Elwood is truly a scientific citizen of the world. He is well-known in academic centers throughout the world and has trained many independent investigators who now hold academic positions in institutions around the world. Like many other distinguished scientists he has made important contributions to numerous scientific organizations and has held positions of responsibility in assemblies, councils, advisory boards and committees, too numerous to list. His scientific contributions have been recognized by a large number of prestigeous awards, richly deserved. However to those who have had the pleasure to work with him and know him personally, his gentle humanity, honesty and consideration of others also distinguishes him as an outstanding friend and associate. His literary talent and sense of humor are also appreciated. Many a meeting has been enlived and stimulated by one of his scientific quatrains. One of these "classics" was penned to clarify the need for ER assays to guide the surgeon planning ablative surgery for breast cancer (Jensen et al., 1967b):

> The surgeon who strives for perfection
> Needs some basis for patient selection.
> He would like to be sure
> There's a good chance of cure,
> Before he begins the resection.

It is clear that his scientific contributions will live on through the scientific literature. But it is also certain that his outstanding human attributes will long be remembered by the large group of his scientific friends, associates and admirers.

REFERENCES

DeSombre ER, Mohla S, Jensen EV (1975). Receptor transformation, the key to estrogen action. J Steroid Biochem 6:469–473.

DeSombre ER, Carbone PP, Jensen EV, McGuire WL, Wells, Jr. SA, Wittliff JL, Lipsett MB (1979). Steroid receptors in breast cancer. New Engl J Med 301:1011–1012.

DeSombre ER, Jensen EV (1980). Estrophilin assays in breast cancer: Quantitative features and application to the mastectomy specimen. Cancer 46:2783-2788.

Gorell TA, DeSombre ER, Jensen EV (1977). Purification of nuclear estrogen receptors. In James VHT (ed): Proc 5th Internatl Cong Endocrinol, Vol 1 Amsterdam: Excerpta Medica Found, pp 467-472.

Gorski J, Toft D, Shyamala G, Smith D, Notides A (1968). Hormone receptors: Studies on the interaction of estrogen with the uterus. Recent Prog Hormone Res 24:45-80.

Greene GL (1984). Application of immunochemical techniques to the analysis of estrogen receptor structure and function. In Litwack G (ed): "Biochemical Actions of the Hormones," Vol XI New York: Academic Press, pp 207-239.

Greene GL, Sobel NB, King WJ, Jensen EV (1984). Immunochemical studies of estrogen receptor. J Steroid Biochem 20:51-56.

Greene GL, Gilna P, Waterfield M, Baker A, Hort Y, Shine J (1986). Sequence and expression of human estrogen receptor complementary DNA. Science 231:1150-1154.

Huggins C, Bergenstal DM (1952). Inhibition of human mammary and prostatic cancers by adrenalectomy. Cancer 12:134-141.

Huggins C, Jensen EV (1954). Significance of the hydroxyl groups of steroids in promoting growth. J Exp Med 100:241-246.

Huggins C, Jensen EV (1955). The depression of estrone-induced uterine growth by phenolic estrogens with oxygenated functions at positions 6 and 16: The impeded estrogens. J Exp Med 102:335-346.

Huggins C, Jensen EV, Cleveland AS (1954). Chemical structure of steroids in relation to promotion of growth of the vagina and uterus of the hypophysectomized rat. J Exp Med 100:225-240.

Jacobson HI, Gupta GN, Fernandez C, Hennix S, Jensen EV (1960). Determination of tritium in biological material. Arch Biochem Biophys 86:89-93.

Jensen EV (1965). Mechanism of estrogen action in relation to carcinogenesis. Canad Cancer Conf 6:143-165.

Jensen EV (1967). Estrogen receptor: Ambiguities in the use of this term. Science 159:1261.

Jensen EV, DeSombre ER (1972). Estrogens and progestins. In Litwack G (ed): "Biochemical Actions of Hormones". New

York: Academic Press, pp 215-255.

Jensen EV, DeSombre ER (1973). Estrogen-receptor interaction. Estrogenic hormones effect transformation of specific receptor porteins to a biochemically functional form. Science 182:126-134.

Jensen EV, Jacobson HI (1960). Fate of steroid estrogens in target tissues. In Pincus G, Vollmer EP (eds): "Biological Activities of Steroids in Relation to Cancer," New York: Academic Press, pp 161-174.

Jensen EV, Jacobson HI (1962). Basic guides to the mechanism of estrogen action. Recent Prog Hormone Res 18:387-414.

Jensen EV, Jacobson HI, Flesher JW, Saha NN, Gupta GN, Smith S, Colucci V, Shiplacoff D, Neumann HG, DeSombre ER, Jungblut PW (1966). Estrogen receptors in target tissues. In Pincus G, Nakao T, and Tait JF (eds): "Steroid Dynamics," New York: Academic Press, pp 133-156.

Jensen EV, DeSombre ER, Jungblut PW (1967a). Interaction of estrogens with receptor sites in vivo and in vitro. Proc. 2nd Internat Congr Hormonal Steroids, Milan 1966. Amsterdam: Exerpta Medica Foundation pp 492-500.

Jensen EV, DeSombre ER, Jungblut PW (1967b). Estrogen receptors in hormone-responsive tissues and tumors. In Wissler RW, Dao T, Wood, S Jr (eds): "Endogenous Factors Influencing Host-Tumor Balance", Chicago: University of Chicago Press, pp 15-30, 68.

Jensen EV, Hurst DJ, DeSombre ER, Jungblut PW (1967c). Sulfhydryl groups and estradiol-receptor interaction. Science 158:385-387.

Jensen EV, Suzuki T, Kawashima T, Stumpf WE, Jungblut PW, DeSombre ER (1968). A two step mechanism for the interaction of estradiol with rat uterus. Proc Natl Acad Sci USA 59:632-638.

Jensen EV, Block GE, Smith S, Kyser K, DeSombre ER (1971a). Estrogen receptors and breast cancer response to adrenalectomy. Nat Cancer Inst Monogr 34: 55-70.

Jensen EV, Numata M, Brecher PI, DeSombre ER (1971b). Hormone-receptor interaction as a guide to biochemical mechanism. Biochem Soc Symp 32:133-159.

Jensen EV, Jacobson HI, Smith S, Jungblut PW, DeSombre ER (1972). The use of estrogen antagonists in hormone receptor studies. Gynec Invest 3:108-122.

Jensen EV, Smith S, DeSombre ER (1976). Hormone dependency in breast cancer. J Steroid Biochem 7:911-917.

Jensen EV, Greene GL, Closs LE, DeSombre ER, Nadji M (1982). Receptors reconsidered: A 20-year perspective. Recent Progress Hormone Res 38:1-40.

Jordan VC, Jacobson HI, Keenan EJ (1986). Determination of estrogen receptor in breast cancer using monoclonal antibody technology: results of a multicenter study in the United States. Cancer Res 46:4237s-4240s.

Jungblut PW, DeSombre ER, Jensen EV (1967). Estrogen receptors in induced rat mammary tumor. Gummel H, Kraatz H, and Bacigalupo (eds): "Hormone in Genese und Therapie des Mammacarcinoms", Abhandlungen der deutschen Akademie der Wissenschaften zu Berlin, Berlin: Akademie-Verlag, pp109-123.

Jungblut PW, Hatzel I, DeSombre ER, Jensen EV (1967). Uber Hormon-"Receptoren": Die oestrogenbindenden Prinzipien der Erfolgsorgane. Colloq Ges Physiol Chem 18:58-86.

King WJ, Greene GL (1984). Monoclonal antibodies localize oestrogen receptor in the nuclei of target cells. Nature (Lon) 307:745-747.

King WJ, DeSombre ER, Jensen EV, Greene GL (1985). Comparison of immunocytochemical and steroid-binding assays for estrogen receptor in human breast tumors. Cancer Res 45:293-304.

Knight WA, Livingston RB, Gregory EJ, McGuire WL (1977). Estrogen receptor as an independent prognostic factor for early recurrence in breast cancer. Cancer Res 37:4660-4671.

Leclercq G, Bojar H, Goussard J, Nicholson RI, Pichon M-F, Piffanelli A, Pousette A, Thorpe S, Lonsdorfer M (1986). Abbott monoclonal enzyme immunoassay measurement of estrogen receptors in human breast cancer: A European Multicenter Study. Cancer Res 46: 4233s-4236s.

McGuire WL, Carbone PP, Vollmer EP, (eds) (1975). "Estrogen Receptors in Human Breast Cancer." New York: Raven Press.

Mohla S, DeSombre ER, Jensen EV (1972). Tissue-specific stimulation of RNA synthesis by transformed estradiol-receptor complex. Biochem Biophys Res Commun 46: 661-667.

Toft D, Gorski J (1966). A receptor molecule for estrogens: Isolation from the rat uterus and preliminary characterization. Proc Natl Acad Sci USA 55:1574-1581.

Walter P, Green S, Greene G, Krust A, Bornert J-M, Jeltsch J-M, Staub A, Jensen E, Scrace G, Waterfield M, Chambon P (1985). Cloning of the human estrogen receptor cDNA. Proc Natl Acad Sci USA 82:7889-7893.

I. STEROID RECEPTORS: STRUCTURE AND FUNCTION

Molecular Endocrinology and Steroid
Hormone Action, pages 33–40
© 1990 Alan R. Liss, Inc.

STRUCTURE, FUNCTION AND IMMUNOLOCALIZATION OF RABBIT AND
HUMAN PROGESTERONE RECEPTORS

M. MISRAHI, H. LOOSFELT, M. ATGER, A. BAILLY, M.
PERROT-APPLANAT, F. LOGEAT, A. GUIOCHON-MANTEL, F.
LORENZO AND E. MILGROM.

Hormones et Reproduction (INSERM U. 135), Faculté de
Médecine Paris-Sud, 94275 Le Kremlin-Bicêtre Cedex France

INTRODUCTION

Progesterone receptors were initially characterized at
the beginning of the 1970's (Milgrom and Baulieu, 1970) .
Since this period their study has relied entirely on
their steroid binding properties. Recently, the
availability of antibodies and, especially monoclonal
antibodies (Logeat et al., 1983), against the receptor,
as well as the cloning of the cDNA and of the gene, have
opened new possibilities in the study of these important
regulatory proteins.

Cloning of rabbit and human PR cDNAs: The rabbit
progesterone receptor cDNA was cloned using the
expression vector λgt 11 (Loosfelt et al., 1986). Poly A
+ RNAs from uteri of estrogen-treated rabbits were
enriched for PR mRNA by sucrose gradient
ultracentrifugation. Random primed cDNAs were synthesized
and inserted into λgt 11 DNA. A library of 3 X 10⁶ clones
was established and screened with both monoclonal (Logeat
et al., 1983) and polyclonal anti PR antibodies. Two
clones reacted with both antibodies. These clones were
shown to encode PR since they hybridized with a mRNA
species only present in target tissues whose
concentration was increased by estrogen treatment.
Moreover, the fusion protein synthesized by these clones
inhibited the binding of the monoclonal antibody to PR
and conversely, purified PR inhibited the binding of the
monoclonal antibody to the fusion protein. These clones

were then used to screen a λgt 10 library which contained large cDNA inserts. Several overlapping clones were isolated allowing the determination of the sequence of the complete coding region of rPR mRNA (Loosfelt et al., 1986). Two rPR mRNA species of 5.9 and 6.5 Kb were detected on Northern blots. They were shown to differ by the size of their 3' non coding sequence. Sequence analysis of the 3' non coding region revealed the presence of three different polyadenylation signals (3058-3553 bp after the TGA stop codon) (Misrahi et al., 1988). Two messenger species were more abundant than the third one.

In order to clone the cDNA for human PR, we established a λgt 10 library from human breast cancer cell line T47D messenger RNAs. The library was screened with probes corresponding to the 5', middle and 3' region of rPR cDNA. Four overlapping clones were used to establish the full coding sequence of hPR mRNA (Misrahi et al.,1987). Northern blot analysis with hPR cDNAs revealed a similar pattern in T47D and MCF7 breast cancer cell lines and in human uterine tissues with four main bands at ~ 5100, 4300, 3700 and 2900 nucleotides and weaker bands at ~ 5900 and ~ 11000. The concentration of PR mRNA displayed the same pattern as that observed for the protein.

Analysis of amino acid sequences deduced from rPR and hPR mRNAs showed proteins of 930 and 933 AA, respectively with calculated molecular weights of 98554 and 98868 Da and with a 88.3 % overall homology. The central basic cysteine-rich region involved in DNA binding was 100 % conserved in both receptors (positions 568 - 645 and 567 -644 respectively) and exhibited extensive homology with other members of the steroid-thyroid hormone receptor family. The hydrophobic C-terminal part involved in steroid binding displayed a single amino acid change between hPR and rPR. It showed a significant degree of conservation with other steroid hormone receptors, especially the human glucocorticoid and mineralocorticoid receptors, but poor homology with thyroid and retinoic acid receptors. The hinge region located between the DNA and the steroid binding domains had only 75 % homology between hPR and rPR; four insertions were found in hPR. A weak homology could be detected only with the corresponding domain of hGR. In the N-terminal part which is the most divergent, the two regions of highest homology between rPR and hPR are found which may possibly have a functional role in the interaction with the transcriptional machinery. This N-terminal region of the PR is very rich in proline and is thus probably very flexible. The acidic character is similar to that of factors involved in transcriptional control (Hope et al., 1988). The N-terminal part of the receptor has been shown to be the only immunogenic region and can be subdivided into 4 domains which are recognized by different groups of monoclonal antibodies (Lorenzo et al., 1988). This N-terminal part of PR is very different in size and amino acid sequence from that present in other members of the steroid thyroid hormone receptor family.

PR mRNA heterogeneity and origin of the A and B forms of progesterone receptor in rabbit and human cells (Misrahi et al., 1988):
Analysis of rabbit and human PR mRNAs using short probes, corresponding to the beginning and to the end of the coding region, shows similar patterns thus suggesting the absence of different mRNA species coding for the A and B forms of receptor (the A form is derived from the B form by deletion of its N-terminus (Lorenzo et al., 1988). Transcription translation experiments using rabbit and human PR cDNAs yielded a major band of apparent molecular weight ~ 110 kDa (corresponding to the B form) and minor shorter bands. These bands corresponded to abortive peptides lacking various lengths of the C-terminus since they were immunoprecipitated by a monoclonal antibody which recognizes an epitope at the very beginning of the protein (between amino acids 1 to 60). Thus, at least in vitro, the two forms of the PR are not derived by use of different translation initiation sites on the same messenger RNA.

Chromosomal localization of the human progesterone receptor gene (Rousseau-Merck et al. 1987):
hPR gene was mapped by in situ hybridization to chromosome 11 band q22-q23 using two cDNA probes corresponding to the 5' and 3' part of the coding sequence. The nearest known neighbour is the c-ets1 oncogene.

Analysis of the promoter and 5' flanking region of the rabbit progesterone receptor gene (Misrahi et al., 1988):

rPR transcription start sites were determined using primer extension experiments and S1 mapping. They are located at adenines present 699 and 712 bp upstream from the first ATG of the open reading frame. The 5' non coding region is thus very long. It is also very G + C rich (63.8%) and contains three short open reading frames. The sequence of 2761 bp upstream from the first initiation site was determined. The promoter region is characterized by the absence of a typical TATA BOX (a TAGAAA motif is present at - 17 and a TAGA at - 37 bp). A CAACT sequence is found at position -100. One consensus binding site for Sp1 is found at position -51. PR level is enhanced by estradiol and down regulated by progesterone. Scanning the 5' flanking region for consensus sequences of hormone responsive elements allowed the location of putative estradiol and progesterone responsive elements to be deduced. Transfection and receptor binding studies are now being performed to determine their functional significance. Consensus sequences for other known regulatory elements were also found (SV40 core enhancer, heat shock element, silencer element). A repetitive sequence of the C family, similar to the one present in the upstream region of the progesterone-regulated uteroglobin gene is found at approximately the same distance (- 2.5 Kb).

Phosphorylation of the progesterone receptor (Logeat et al., 1985):
PR undergoes a basal phosphorylation and is further phosphorylated when hormone is administered. The latter phosphorylation may be visualised by an "upshift" in its migration on SDS polyacrylamide gel electrophoresis. The role of the phosphorylation in the mechanism of regulation of target genes is not yet understood. Phosphorylation may be also involved in receptor "down regulation" in response to the hormone (Vu Hai et al., 1977). A protein kinase which copurifies with the receptor has been identified but its action is independent of hormone administration in vitro and does not provoke the "upshift" in receptor electrophoretic migration (Logeat al., 1987).

Interaction of PR with DNA:
Direct _in vitro_ interactions of the progesterone receptor with specific DNA sequences of the uteroglobin gene have been analysed (Bailly et al., 1986). Two groups of high affinity binding sites have been mapped in the 5' flanking region, up to -2.7 Kb, of the gene. However, the interaction of PR with its target sequences seems to differ _in vivo_ and _in vitro_. In vivo the hormone is necessary for the interaction while it is dispensible _in vitro_: purified PR has the same affinity for hormone responsive elements in its free state, complexed with the hormone or with a hormonal antagonist (the RU 38486), and after hormone dependent phosphorylation _in vivo_. This discrepancy could be explained by the presence _in vivo_ of an inhibitor, stabilizing the PR in its non active form. The hormone would allow this inhibitor to be displaced by a conformational change of the complex, yielding the activated form of the receptor. The purification process could have eliminated this factor and thus provoked spontaneous activation of the receptor.

In nearly all hormonally regulated genes, several receptor binding sites have been observed. These sites are very distant from each other in the uteroglobin gene. This raises the question of the possibility of an interaction between the different receptor molecules bound to DNA. Using electron microscopy, direct interactions between two oligomers of receptor bound at two distinct specific DNA sites of the uteroglobin gene have been visualised (Theveny et al, 1987). These interactions lead to the generation of DNA loops. This mechanism could be important for receptor binding cooperativity and for the interaction with other transcriptional factors.

Immunocytochemical studies of progesterone receptor:
Using specific monoclonal antibodies, PR molecules have been located almost exclusively in the nuclei of target cells in immature rabbits or castrated guinea pigs by the immunoperoxidase method (Perrot-Applanat et al., 1985). Heterogeneity in the staining was observed between cells of the same type in several tissues suggesting a difference in hormonal sensitivity or a variation in PR content during the cell cycle. To analyze subcellular localization of PR, electron microscopy studies were

performed with an immunogold method (Perrot-Applanat et al., 1986). Most of the PR was associated in the nucleus with condensed chromatin but its distribution varied according to the hormonal status. After administration of progesterone, the receptor was detected at the border of condensed chromatin and nucleoplasm and to a lesser extent over dispersed chromatin in the nucleoplasm. These sites have been shown to be the most active regions of gene transcription. A very small fraction of PR was detected in the cytoplasm where it was associated with the rough endoplasmic reticulum and free ribosomes. It could represent either the newly synthesized PR or indicate a function of the PR at the translational level. Monoclonal anti PR antibodies were used to study breast cancer biopsies (Perrot-Applanat et al., 1987). Estradiol and progesterone receptor determination were performed on adjacent frozen sections of the same biopsy. There was a good correlation with the steroid binding assay of PR and ER levels in these tumors.

The immunoperoxidase method was used to study PR and ER variations in the endometrium during the normal menstrual cycle (Garcia et al., 1988). PR variations during the late luteal phase were shown to be very useful in the assessment of the cumulative effect of progesterone during the second part of the cycle (Bayard et al., 1978 ; Kreitmann et al., 1979). This method may be of interest in the evaluation of infertility in women.

ACKNOWLEDGEMENTS

We thank Valérie Francois for typing this manuscript.
This work was supported by the Institut National de la Santé et de la Recherche Médicale (INSERM), the Centre National de la Recherche Scientifique (CNRS), the Foundation pour la Recherche Médicale Française (FRMF), the Association pour la Recherche sur le Cancer (ARC) and the Unité d'Enseignement et de Recherche Kremlin-Bicêetre.

REFERENCES

Bailly A, Le Page C, Rauch M, Milgrom E (1986). Sequence-Specific DNA binding of the progesterone receptor to the uteroglobin gene: effects of hormone, antihormone and receptor phosphorylation. The EMBO J 5:3235-3241.

Bayard F, Damilano S, Robel P, Baulieu EE (1978). Cytoplasmic and nuclear estradiol and progesterone receptors in human endometrium. J Clin Endocrinol Metab. 46:635-648.

Garcia E, Bouchard P, De Brux J, Berdah J, Frydman R, Schaison G, Milgrom E, Perrot-Applanat M (1988). Use of immunocytochemistry of progesterone and estrogen receptors for endometrial dating. J Clin Endocrinol Metab. In press.

Hope IA, Mahadevan S, Struhl K (1988). Structural and functional characterization of the short acidic transcriptional activation region of yeast GCN4 protein. Nature. 333:635-640.

Kreitmann B, Bugat R, Bayard F (1979). Estrogen and progestin regulation of the progesterone receptor concentration in human endometrium. J Clin Endocrinol Metab. 49-926.

Logeat F, Vu Hai MT, Fournier A, Legrain P, Buttin G, Milgrom E (1983). Monoclonal antibodies to rabbit progesterone receptor. Cross-reaction with other mammalian progesterone receptors. Proc Natl Acad Sci USA 80:6456-6459.

Logeat F, Le Cunff M, Pamphile R, Milgrom E (1985). The nuclear-bound form of the progesterone receptor is generated through a hormone dependent phosphorylation. Biochem Biophys Res Commun. 131:421-427.

Logeat F, Le Cunff M, Rauch M, Brailly S, Milgrom E (1987). Characterization of a casein kinase which interacts with the rabbit progesterone receptor. Differences with the in vivo hormone-dependent phosphorylation. Eur J Biochem. 170: 51-57.

Loosfelt H, Atger M, Misrahi M, Guiochon-Mantel A, Mériel C, Logeat F, Bénarous R, Milgrom E (1986). Cloning and sequence analysis of rabbit progesterone-receptor complementary DNA. Proc Natl Acad Sci 83:9045-9049.

Lorenzo F, Jolivet A, Loosfelt H, Vu Hai MT, Brailly S, Perrot-Applanat M, Milgrom E (1988). A rapid method of epitope mapping. Application to the study of immunogenic domains and to the characterization of various "forms" of rabbit progesterone receptor. Eur J Biochem. In press.

Milgrom E, Baulieu E E (1970). Progesterone in uterus and plasma. I - Binding in rat uterus 105.000 g supernatant. Endocrinology 87:276-287.

Misrahi M, Atger M, D'Auriol L, Loosfelt H, Meriel C, Fridlansky F, Guiochon-Mantel A, Galibert F, Milgrom E (1987). Complete amino-acid sequence of the human progesterone receptor deduced from cloned cDNA. Biochem Biophys Res Commun 143:740-748.

Misrahi M, Loosfelt H, Atger M, Meriel C, Zerah V, Dessen P, Milgrom E (1988). Organization of the entire rabbit progesterone receptor mRNA and of the promoter and 5'flanking region of the gene. Nucleic Acids Research. 16:5459-5472.

Perrot-Applanat M, Logeat F, Groyer-Picard MT, Milgrom E (1985). Immunocytochemical study of mammalian progesterone receptor using monoclonal antibodies. Endocrinology 116:1473-1484.

Perrot-Applanat M, Groyer-Picard MT, Logeat F, Milgrom E (1986). Ultrastructural localization of the progesterone receptor by an immunogold method: Effect of hormone administration. J Cell Biol 102:1191-1199.

Perrot-Applanat M, Groyer-Picard MT, Lorenzo F, Vu Hai MT, Pallud C, Spyratos F, Milgrom E (1987). Immunocytochemical study with monoclonal antibodies to progesterone receptor in human breast tumors. Cancer Res 47:2652-2661.

Rousseau-Merck MF, Misrahi M, Loosfelt H, Milgrom E, Berger R (1987). Localization of the human progesterone receptor gene to chromosome 11q22q23. Human Genet 77:280-282.

Theveny B, Bailly A, Rauch C, Rauch M, Delain E, Milgrom E (1987). Association of DNA-bound progesterone receptors. Nature. 329:79-81

Vu Hai MT, Logeat F, Warembourg M, Milgrom E (1977). Hormonal control of progesterone receptors. Ann. N.Y. Acad. Sci. 286:199-209.

Molecular Endocrinology and Steroid
Hormone Action, pages 41–52
© 1990 Alan R. Liss, Inc.

HUMAN PROGESTERONE RECEPTORS: SYNTHESIS, STRUCTURE, AND
PHOSPHORYLATION

Kathryn B. Horwitz, Philip L. Sheridan,
Lisa L. Wei, and Nancy L. Krett

Departments of Medicine (KBH, LLW, NLK) and
Pathology (KBH, PLS), University of Colorado
Health Sciences Center, Denver, Colorado 80262

The use of progesterone receptor (PR) measurements and
progestins in the treatment of breast cancer (Horwitz et al.,
1985) and the use of progesterone antagonists in contraception
(Herrmann et al., 1982) has required an understanding of the
actions of these hormones at a basic molecular level. To study
the structure and function of human progesterone receptors (hPR)
we have used $T47D_{co}$, a PR-rich human breast cancer cell line
(Horwitz et al., 1982), and in situ photoaffinity labeling (Horwitz
et al., 1983; Horwitz et al., 1985); we have also purified the PR
from T47D cells and have made anti-PR monoclonal antibodies
(Estes et al., 1987). This chapter describes some of our recent
studies using these reagents (Wei et al., 1987) to analyze the
structure of hPR.

PHOTOAFFINITY LABELING AND THE ORIGIN OF THE A- AND
B-PROTEINS

In situ photoaffinity labeling uses ultraviolet irradiation to
covalently link the radioactive progestin R5020 (17,21-dimethyl-
19-nor-pregn-4,9-diene-3,20-dione; [17α-methyl-^{3}H]) to receptors
in intact cells. Briefly, cells either attached to plastic, or
harvested but still intact, are incubated with [^{3}H]R5020 for 2-
4 hrs at 0°C to keep PR in the untransformed cytosolic state,
or for 5-10 min at 37°C to transform PR to the tight nuclear
binding state. The intact cells are then irradiated with UV for
2 min. The cells can be solubilized directly; or they can be
homogenized and then receptors from various subcellular
compartments, including cytosols, nuclear extracts, microsomal
pellets and residual nuclear pellets, can be solubilized and

denatured. The denatured radioactive receptors are separated by SDS–polyacrylamide gel electrophoresis and visualized by fluorography. Alternatively after transfer to nitrocellulose, the proteins can be analyzed by combined immunoblotting and fluorography (Wei et al., 1987). In situ photolabeled untransformed hPR are composed of 2 major hormone binding species. The heavier B protein is a doublet or triplet of M_r 117–120,000 daltons, and the smaller A protein is a singlet of M_r 94,000 daltons.

These proteins are remarkably stable. Even incubation for 1 hour at 37°C fails to degrade the B protein either to A or to any other fragments. We have previously shown that proteins A and B bind only progestins specifically (Horwitz et al., 1985), and that they are present in equimolar amounts (Horwitz et al., 1985). They can be demonstrated even if homogenization and centrifugation are eliminated -- that is, by in situ photoaffinity labeling of intact cells, solubilization of the cells in buffer containing detergent plus protease inhibitors, denaturation, and immediate electrophoresis (Horwitz et al., 1985). The "doublet B - singlet A" is characteristic not only of activated and untransformed receptors, having low affinity for nuclei, but also of transformed receptors shortly after they have acquired tight chromatin binding capacity as a result of hormone treatment. Thus, equimolar amounts of B- and A-proteins are found whether receptors are soluble in cytosols or extracted from nuclei.

The argument has been made that A is formed in vitro as an artifactual proteolytic fragment of B; this argument is supported by several studies. First is the fact that despite repeated demonstrations of two progestin-binding proteins in chick oviduct and human breast cancer cells, two hormone-binding proteins are usually not demonstrated for the other steroid receptors (Horwitz et al., 1985). On occasions when two proteins are seen, the proteolysis explanation is invoked (Loosfelt et al., 1984). Second are studies demonstrating structural similarity between chick or human A- and B-proteins by peptide mapping of photolabeled receptors (Horwitz et al., 1985; Birnbaumer et al., 1983; Gronemeyer et al., 1983); and by immunoreactivity (Gronemeyer et al., 1985). These are compelling findings but are also inconclusive, since they fail to reveal the ca. 25 kDa fragment that must be generated from B but not A if the proteolysis argument is true. Nor do the studies prove a proteolysis mechanism, since other explanations for the biosynthesis of partially homologous proteins are possible (see below). Third are immunoblot studies of rabbit uterine PR.

These studies show that if homogenization is done quickly, little or no protein A is generated. Unfortunately, the antibody used for these experiments has not been fully characterized — receptor purification data are unpublished to our knowledge (Logeat et al., 1983; Logeat et al., 1985), and the antibody used for the blotting studies failed to bind to 50% of native PR in salt-containing sucrose gradients (Logeat, 1983), suggesting that it may have very different affinities for proteins A and B, and that it belongs to the class of B-specific antibodies commonly generated from partially purified PR (Estes et al., 1987). Thus, the protease hypothesis has not, in our opinion, been conclusively established. A cocktail of protease inhibitors is routinely added to homogenization buffers in our studies, but A-receptor levels are unaffected.

Our studies of breast cancer hPR suggest that protein B (M_r 120,000) and protein A (M_r 94,000) are integral intracellular proteins and that A is not generated by proteolysis: (a) Regardless of the site from which they are extracted (cytoplasmic or nuclear), proteins A and B are found in equimolar amounts (Horwitz et al., 1985; Horwitz et al., 1983). If a protease is degrading protein B, it must be present in cytosols as well as in nuclear extracts, it must be equally active in buffers of different ionic strengths and pH, and its activity must stop after degrading half of the B molecules. (b) Degradation of B to A cannot be demonstrated during in vitro incubations forcing the conclusion that the putative enzyme acts instantaneously on half of the B molecules when the cells are first broken. (c) Both receptor proteins are seen in cells photolabeled in situ, then lysed directly in buffer containing detergent and protease inhibitors, and immediately subjected to electrophoresis (Horwitz et al., 1985). This suggests that the putative protease acts intracellularly (Birnbaumer et al., 1983). Furthermore, we have shown that the two receptor proteins are dissimilarly modified in their untransformed state. Since B is a doublet or triplet while A is a singlet in holoreceptors, one would have to postulate that if A is a proteolytic artifact, then the offending protease clips out a domain of B containing the site(s) responsible for this heterogeneity. However, such a protease cannot be invoked for the 30 minute resident nuclear receptors, where protein B is a singlet while protein A is a doublet (Horwitz et al., 1985). Here different proteases would have to be involved, that clip B at two sites 3000 daltons apart to generate the A doublet. Thus, we are left to conclude either that B is subject to degradation by a series of unusual intracellular proteases, or that A is a protein closely related to B, that is formed intracellularly by other mechanisms. Since antigenic and hormone binding peptides below the size of the A-protein are similar for both A- and B-receptors, since B and A bind hormone (a site at the C-terminal end of the molecule) and DNA (in the center of the molecule), and since A/B-specific and B-specific antibodies exist, but no A-specific antibodies have been described, we postulate that B-receptors differ from A by having additional N-terminal sequences. Possible mechanisms for the synthesis of such partially homologous proteins include gene duplication, alternate transcription from a single gene, synthesis of multiple messages by alternate processing of a single precursor RNA, or use of alternate translation start sites from a single message. Two putative in-frame AUG codons, satisfying the Kozak rules (Kozak, 1986) for translation initiation sites, are present 165 bases apart in the hPR message (Misrahi et al., 1987). These could theoretically code for two homologous proteins with the shorter protein truncated at the N-terminal end and -18,000 daltons smaller. We have preliminary evidence that translation initiation from the downstream AUG leads to synthesis of A-receptors.

HUMAN PR PURIFICATION AND MONOCLONAL ANTIBODY PRODUCTION

In order to address the questions regarding the origins of the A- and B-proteins, we required antibodies specific to human PR. To this end we purified B-receptors from T47D cells by immunoaffinity chromatography using an antibody designated PR-6 developed by D. Toft and colleagues (Sullivan et al., 1986) that binds the B-receptors of chick oviducts and cross-reacts with human B-proteins. Single step immunoaffinity chromatography resulted in enrichment of human B-receptors (identified by immunoblotting with PR-6 and by photoaffinity labeling with [^{3}H]R5020) from a specific activity of 0.71 to 1,915 pmol/mg protein (or 23% purity) with 27% yield (Table 1). Purity and yields as judged by gel electrophoresis and densitometric scanning of the silver-stained B-protein were approximately 1.7 fold higher due to partial loss of hormone binding activity at the elution step. A second purification step using DEAE chromatography gave further enrichment to 3,720 pmol/mg protein (or 44% purity) to yield only B-receptors plus one other component — a 72 kDa protein present in approximately equivalent amounts.

Based on physiochemical properties, single step immuno-affinity purified B-receptors were in the native transformed state:

isolated receptors were maintained as undegraded 120 kDa doublets; they retained their hormone binding activity; and they displayed the correct steroid specificity for PR. Isolated B-receptors also bound efficiently to DNA-cellulose requiring 0.25M salt for their release. They sedimented as 4S monomers on salt-containing sucrose gradients and as a 6S peak in the absence of salt. All these confirm their transformed state. In addition, under these conditions, purified B-receptors were free of the 90 kDa receptor-associated heat shock protein that is always observed to copurify with 8S untransformed receptors in other systems, and they were free also of the nonhormone binding 108 kDa antigen that copurifies with chick oviduct PR. However, in addition to the 120 kDa B-doublet, three silver stained bands are detected in the single-step purified preparations at 72, 62, and 58 kDa. These proteins are not reactive with PR-6 by immunoblotting, nor do they bind [^{3}H]R5020. Thus, they are not receptor fragments but unrelated proteins. The two smaller proteins are abundant cellular proteins that bind the immunomatrix nonspecifically, but the 72 kDa protein is receptor-associated since it cannot be purified from receptor-depleted cytosols. It may be similar to the GR-associated 72 kDa protein described by Gustaffson et al. (Gustaffson et al., 1986) which is a heat shock protein (Welch et al., 1985).

Table 1: Purification of Human B Receptors from T47D Cells by Single Step Immunoaffinity Chromatography

	Total Protein		B Receptors			
	(mg/ml)	(mg)	pmol[a]	yield (%)	sp. act. (pmol/mg)	purity[b] (%)
Cytosol	14.40	1,152.0	814	---	0.71	
pH eluate	0.021	0.132	221[c]	27.1	1,915.00	23

[a]   B receptors in starting cytosol (estimated to be 50% of total receptors) were measured by steroid binding assay.

[b]   Theoretical purity for B receptors based on M_r of 120,000 = 8,352 pmol/mg protein.

[c]   Receptor-hormone binding of the purified product was measured by hydroxylapatite assay with inclusion of 1% carrier albumin. Nonspecific binding was determined with carrier albumin alone.

Values are average determinations from three separate purifications.

The partially purified B-receptors from T47D cells were used to immunize mice, and three monoclonal antibodies (MAbs) against human PR were produced. They have been subcloned and are clonally stable. Their identifying codes and other properties are summarized in Table 2, and the monospecificity of the IgGs has been shown by immunoblotting. Although mice were injected only with B-receptors, the production of one antibody (AB-52) that cross-reacts with both A- and B-receptors again attests to the structural homology of these two proteins. The three IgGs are capable of shifting labeled PR on sucrose gradients and demonstrating PR in cell and tumor samples by immunohistochemistry and flow cytometry. We have not yet determined whether B-30 and B-64 bind the same or different epitopes unique to B; clearly AB-52 binds a region common to both A- and B-proteins. The development of the monoclonals has been described in detail elsewhere (Estes et al., 1987).

Table 2. Monoclonal Antibodies to Human PR

MAb	Antibody Subtype	Reactivity[1] A	B
B-30	IgG$_1$	−	+
B-64	IgG$_1$	−	+
AB-52	IgG$_1$	+	+

[1]   Reactivity with 94 kDa A- or 120 kDa B-receptor was determined by immunoblotting. See Estes et al. 1987 for details.

IMMUNE ANALYSIS OF hPR STRUCTURE

Because it is B-specific and binds to native human PR, PR-6 was used to study the nature of the association between the A- and B-proteins in the untransformed 8S state and in the transformed 4S state (Wei et al., 1987). This is of interest since there are conflicting models for the molecular interaction of the A- and B-proteins. One model holds that A and B dimerize and that they are subunits of a larger holoreceptor (Schrader et al., 1981); the other model holds that A and B exist as separate 8S molecules (Dougherty et al., 1982; Renoir et al., 1984).

Using PR-6 to immunoprecipitate B-receptors, we have been unable to co-precipitate A-receptors even in the presence of sodium molybdate. This should have been possible if A and B are tightly associated. We therefore tested the association of A and B more extensively using sucrose density gradient analysis and in situ photoaffinity labeling. We reasoned that if the dimeric subunit model is correct, addition of PR-6 to receptors stabilized in the 8S state would shift both A- and B-proteins to the bottom of sucrose density gradients, but that A would not be shifted if the two proteins form independent 8S holoreceptor complexes. A study was performed in which hPR, covalently labeled in situ with [^{3}H]R5020 by UV irradiation, were incubated with PR-6 or a control antibody and then sedimented on sucrose gradients in the continuous presence of molybdate and protease inhibitors to maintain intact and native PR conformation. Aliquots of every gradient fraction were counted to obtain the [^{3}H]R5020-binding profile, and additional alioquots of the bottom fractions and peak 8S fractions were analyzed by gel electrophoresis and fluorography since they were photoaffinity labeled. The control antibody had no effect on 8S sedimentation of hPR, and electrophoretic analysis showed that all B- and A-receptors remained in the 8S peak. In contrast, after addition of PR-6, half of the 8S radioactivity was shifted to heavier sedimenting forms. Electrophoretic analysis showed only B present in the antibody-shifted fraction at the bottom of the gradient while A, separated from B, remained at 8S. We concluded that A and B do not dimerize but that each exists as a separate 8S multimetic receptor complex either because of self-association, or because of association with nonhormone binding proteins. Two such proteins of 90 and 72 kDa co-purify with untransformed human B-receptors (Estes et al., 1987).

PR-6 can also separate B-receptors from A-receptors that have been transformed to the 4S species by treatment with salt. Approximately half of the radioactivity seen in the 4S peak in the presence of control antibody was shifted to heavier aggregates upon addition of PR-6 and a secondary antibody. The control 4S peak contained both A and B, but only A remained at 4S after PR-6 addition while the B-receptors were shifted to heavier sedimenting fractions. It appears then that antibody PR-6 cross-reacts both with the native 8S as well as the transformed 4S forms of B-receptors, and that like the case for 8S, there are two types of 4S species containing either A-protein or B-protein, but not both. The 72 kDa nonhormone binding proatein co-purifies with transformed 4S B-receptors (Estes et al., 1987).

Since A and B are not linked in transformed hPR, the two-stage nuclear binding model in which B is seen as a chromatin-specifying protein that guides A to apropriate DNA binding sites (Schrader et al., 1981), is unlikely to be correct. The alternative model is that A- and B-proteins form independent receptor complexes, and that like other steroid receptors, each binds to DNA. While there appears to be consensus that A-receptors bind DNA, for B-receptors this point has been unsettled (Gronemeyer et al., 1985; Schrader et al., 1981). We have tested the ability of immunopurified transformed human B-receptors to bind DNA-cellulose and find that they do so efficiently (Estes et al., 1987). In unpublished studies we find that PR-negative cells transfected with a cDNA that encodes hPR_A synthesize only A-receptors and can respond to progestin treatment. The laboratories of Chambon (Gronemeyer et al., 1987) and O'Malley (Conneely et al., 1987) have shown that A and B are encoded on a single messenger RNA but are transcribed from alternate initiation sites. The inference is strong that A is a natural form of PR. Tora et al (Tora et al., 1988) demonstrate, furthermore, that A and B may subserve functional differences in transcriptional activation.

Our working model for the structure of native hPR is that there are two classes of untransformed 8S receptors, each associated with nonsteroid binding proteins of 72 kDa and 90 kDa. Transformation results from dissociation of the 90 kDa protein exposing DNA binding sites masked in the 9S receptor forms. Transformed 4S receptors which bind tightly to DNA may be heterodimers composed of one steroid binding protein and the 72 kDa protein. The A and B steroid proteins do not form A-B dimers but exist as separate 8S and 4S molecules (hPR_A and hPR_B) that are independent as DNA binding proteins and function independently to elicit a biological response.

IMMUNE ANALYSIS OF hPR SYNTHESIS AND PHOSPHORYLATION

We have also used the monoclonal antibodies to study hPR synthesis, and the rate of receptor activation to a hormone binding state; to analyze the origin and function of B-receptor heterogeneity; and to analyze hPR phosphorylation.

On immunoblots, purified B-receptors appear as 3 bands of M_r ~114 (B_1), 117 (B_2), and 120 (B_3) kDa. A-receptors are single bands of 94 kDa. To study the origin of B-receptor heterogeneity, T47D cells were pulse labeled with [^{35}S]methionine, then chased with cycloheximide to prevent additional protein synthesis. The labeled proteins were then visualized by immunoprecipitation, gel electrophoresis, blotting, and autoradiography. The data show that B-receptors are synthesized as a single protein of 114 kDa; the B_1 form. Then, while protein synthesis is inhibited, the heavier bands, B_2 and B_3, are generated over a 6 to 10 hour period. Thus, the B-triplet represents a post-translational maturation step. The heavier forms, B_2 and B_3, can be reduced to lighter forms with alkaline phosphatase.

Additional studies show that this slow maturation step is not required for functional activation of the receptors. By using [^{35}S]methionine pulses as short as 10 min, followed by inhibition of protein synthesis, we can show that the 10 min old B_1 singlets and the 10 min old A-receptors are competent to bind hormone and to be transformed to the tight chromatin-bound state. Transformation is demonstrated by the upshift in molecular weight that both B- and A-receptors undergo, after they are bound to chromatin. Thus, in T47D cells, nascent receptors are rapidly activated, and we see no evidence for long-lived proreceptors (unpublished data).

That untransformed hPR are phosphoproteins has been shown using metabolic labeling with [^{32}P]orthophosphate. Again the in situ labeled receptors are analyzed by immunoprecipitation, gel electrophoresis, blotting, and autoradiography. This analysis shows that all 3 bands of the B-receptors, as well as the singlet A-receptors, are phosphorylated in the hormone-free state. Labeling is exclusively on serine residues. After hormone treatment, chromatin-bound PR then undergo a second round of serine phosphorylation that increases the specific activity of [^{32}P] labeling 8-10 fold over untransformed receptors. Reverse phase HPLC analysis of phosphotryptic peptides shows that untransformed A- and B-receptors share five common phosphopeptides, and that a sixth is unique to B. The hormone-dependent nuclear phosphorylation that leads to increased [^{32}P] labeling occurs on at least 3 new peptide residues. To study the possible function of the nuclear phosphorylation reaction, the effects of progestins were compared to those of the anti-progestin RU486 in cells treated with hormones for 0.5 to 48 hrs. The agonists lead to receptor protein processing or down-regulation, with parallel decreases in phosphorylation levels. Like the agonists, the antagonist promotes transformation of PR to the tight chromatin binding state, and nuclear phosphorylation occurs on the same peptides as with the agonists. However, with the antagonist the [^{32}P] labeling intensity is 2-3 fold higher.

Unlike the agonists, RU486 completely inhibits PR processing. Thus, the phosphorylation is not accompanied by down-regulation. We conclude that the multiple forms of B-receptors are due to post-translational phosphorylation, but that this maturation is not required for receptor function, since the nascent receptors are able to bind hormone and be activated. There are at least 5 phosphorylation sites on untransformed A-receptors and 6 on B-receptors, and new sites are phosphorylated after the receptors are bound by hormone. Once in nuclei, PR can be phosphorylated without subsequently being processed, suggesting that phosphorylation affects a step after transformation, but before down-regulation; either the affinity of receptors for DNA or for other factors of the transcription complex (Sheridan et al., 1988).

ACKNOWLEDGMENTS

This chapter is submitted in honor of Elwood Jensen and Jack Gorski. These studies were supported by grants from the National Institutes of Health, the National Science Foundation, and the National Foundation for Cancer Research. They were performed in part in collaboration with Dean Edwards, David Toft, Bert O'Malley, and Julius Gordon.

REFERENCES

Birnbaumer M, Schrader WT, and O'Malley BW (1983). Assessment of structural similarities in chick oviduct progesterone receptor subunits by partial proteolysis of photoaffinity labeled proteins. J Biol Chem 255:1637-1644.

Conneely OM, Maxwell BL, Toft DO, Schrader WT, O'Malley BW (1987). The A and B forms of chicken progesterone receptor arise by alternate initiation of translation of a unique mRNA. Biochem Biophys Res Comm 149:493-501.

Dougherty JJ, Toft DO (1982). Characterization of two 8S forms of chick oviduct progesterone receptor. J Biol Chem 257:3113-3119.

Estes PA, Suba EJ, Lawler-Heavner J, Elashry-Stowers D, Wei LL, Toft DO, Sullivan WP, Horwitz KB, Edwards DP (1987). Immunologic analysis of human breast cancer progesterone receptors. I. Immunoaffinity purification of transformed receptors and production of monoclonal antibodies. Biochemistry 26:6250-6262.

Gronemeyer H, Govindan MV, and Chambon P (1985). Immunologic similarity between the chick oviduct forms A and B. J Biol Chem 260:6916-6925.

Gronemeyer H, Harry P, and Chambon P (1983). Evidence for two structurally related progesterone receptors in chick oviduct cytosol. FEBS Lettr 156:287-292.

Gronemeyer H, Turcotte B, Quirin-Stricker C, Bocquel MT, Meyer ME, Krozowski Z, Jeltsch JM, Lerouge T, Garnier JM, Chambon P (1987). The chicken progesterone receptor: Sequence, expression and functional analysis. The EMBO J 6:3985-3994.

Herrmann W, Wyss R, Provider A (1982). The effects of an antiprogesterone steroid in women; interruption of the menstrual cycle and early pregnancy. CR Acad Sc Paris 294:933-938.

Horwitz KB, Alexander PS (1983). In situ photolinked nuclear progesterone receptors of human breast cancer cells; subunit molecular weights after transformation and translocation. Endocrinology 113:2195-2201.

Horwitz KB, Francis MD, Wei LL (1985). Hormone-dependent covalent modification and processing of human progesterone receptors in the nucleus. DNA 4:451-460.

Horwitz KB, Mockus MB, and Lessey BA (1982). Variant T47D human breast cancer cells with high progesterone receptor levels despite estrogen and antiestrogen resistance. Cell 28:633-642.

Horwitz KB, Wei LL, Sedlacek SM, d'Arville CN (1985). Progestin action and progesterone receptor structure in human breast cancer: A review. Recent Progress in Hormone Research 41:249-316.

Kozak M (1986). Bifunctional messenger RNAs in eukaryotes. Cell 47:481-483.

Logeat F, Hai MTV, Fournier A, Legrain P, Buttin P, Milgrom E (1983). Monoclonal antibodies to rabbit progesterone receptor: Cross-reaction with other mammalian progesterone receptors. Proc Natl Acad Sci USA 80:6456-6459.

Logeat F, LeCunff M, Pamphile R, Milgrom E (1985). The nuclear-bound form of the progesterone receptor is generated through a hormone-dependent phosphorylation. Biochem Biophys Res Comm 131:421-427.

Loosfelt H, Logeat F, Hai MTV, Milgrom E (1984). The rabbit progesterone receptor. Evidence for a single steroid-binding subunit and characterization of receptor mRNA. J Biol Chem 259:14196-14202.

Misrahi M, Atger M, d'Auriol L (1987). Complete amino acid sequence of the human progesterone receptor deduced from cloned cDNA. Biochem Biophys Res Comm 143:740-748.

Renoir JM, Mester J (1984). Chick oviduct progesterone receptor: Stucture, immunology, function. Mol Cell Endocr 37:1-13.

Schrader WT, Birnbaumer ME, Hughes MR, Weigel NL, Grody WW, O'Malley BW (1981). Studies on the structure and function of the chicken progesterone receptor. Rec Prog Horm Res 37:583-633.

Sheridan PL, Krett NL, Gordon JA, Horwitz KB (1988). Human progesterone receptor transformation and nuclear down-regulation are independent of phosphorylation. Molecular Endocrinology, in press.

Sullivan WP, Beito TJ, Proper J, Krco CJ, Toft DO (1986). Preparation of monoclonal antibodies to the avian progesterone receptor. Endocrinology 119:1549-1557.

Tora L, Gronemeyer H, Turcotte B, Gaub M-P, Chambon P (1988). The N-terminal region of the chicken progesterone receptor specifies target gene activation. Nature 333:185-188.

Wei LL, Sheridan PL, Krett NL, Francis MD, Toft DO, Edwards DP, Horwitz KB (1987). Immunologic analysis of human breast cancer progesterone receptors. II. Structure, phosphorylation and processing. Biochemistry 26:6262-6272.

Welch WT, Feramisco JR (1985). Rapid purification of mammalian 70,000-dalton stress proteins: Affinity of proteins for nucleotides. Molec and Cell Biol 5:1229-1237.

Molecular Endocrinology and Steroid
Hormone Action, pages 53–63
© 1990 Alan R. Liss, Inc.

MOLECULAR CLONING AND STRUCTURAL ANALYSIS OF COMPLEMENTARY DNA OF HUMAN AND RAT ANDROGEN RECEPTORS

Chawnshang Chang[1,3], John Kokontis[1], Sarah Swift[1] and Shutsung Liao[1,2]

Ben May Institute[1], Department of Biochemistry and Molecular Biology[2], and Department of Surgery/ Urology[3], University of Chicago, Chicago, Illinois 60637

INTRODUCTION

Molecular cloning of cDNAs for various steroid receptors (SRs) in recent years has made it possible to analyze the structural importance of different receptor domains in binding hormones and in recognition of hormone responsive elements on DNA. While the cloning of these cDNAs was helped greatly by the use of antibodies to these receptors, cloning of cDNA for an androgen receptor (AR) was difficult because monospecific antibodies against AR were not available for screening cDNA libraries. Using oligonucleotide probes and a differential screening method we were able to identify clones with AR-cDNA sequences. These DNA sequences were ligated to construct cDNAs that were able to direct synthesis of full-length human (h) and rat (r) ARs in cell-free systems.

MOLECULAR CLONING

λGT11 cDNA libraries were constructed with polyadenylated RNA from human testis and prostate and also from the ventral prostate of rats. These libraries were initially screened with a 41-b oligonucleotide probe (Chang et al., 1988a) that was highly homologous to nucleotide sequences in the DNA-binding domain of glucocorticoid receptors (GR), estrogen receptors (ER), progesterone receptors (PR), mineralocorticoid receptors (MR), vitamin D receptor (VDR) and thyroid hormone receptor (TR) or the v-erbA oncogene product of avian erythroblastosis virus. The conditions of hybridization during screening were 25%

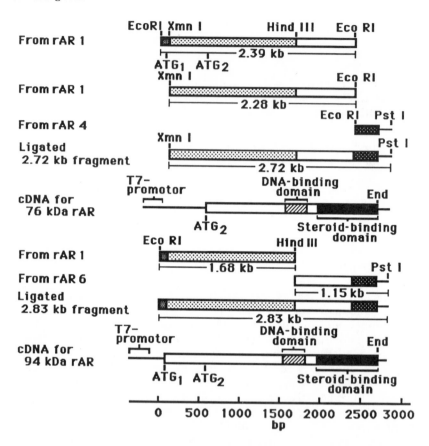

Fig. 1. Strategy used in the construction of cDNA for rat androgen receptor. To obtain the 2.83-kb cDNA that could encode a 94-kDa receptor protein, the 2.39-kb EcoRI/EcoRI cDNA insert of rAR 1 was digested with HindIII to obtain a 1.68-kb fragment. This fragment was ligated to a 1.15-kb DNA fragment that was obtained by digestion of the cDNA insert of rAR 6 clone with HindIII and PstI. The ligated 2.83-kb construct was inserted into a EcoRI and PstI-digested pGEM-3Z vector.

formamide, 5 x Denhardt's solution (0.1% Ficoll 400, 0.1% poly-vinylpyrrolidone, 0.1% bovine serum albumin), 0.1% SDS, 5 x SSC (1 x SSC is 150 mM NaCl, 15 mM sodium citrate), 100 µg/ml denatured salmon sperm DNA, and 1 µg/ml poly(A) at 30° C.

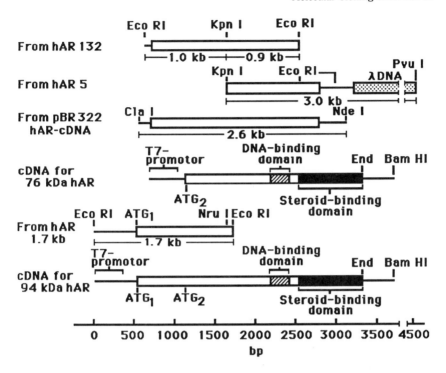

Fig. 2. Strategy used in the construction of cDNA for human androgen receptor. For construction of a full-length hAR cDNA, a 1.7-kb 5' end hAR cDNA fragment was digested with Nru I to give a 1.6-kb fragment that was then ligated to a 2.0-kb hAR fragment obtained by BamHI and NruI digestion of the 2.6-kb hAR in pGEM-3Z (Chang et al., 1988b). The resulting 3.6 kb fragment was ligated into EcoRI and BamHI-digested pGEM-3Z in the same orientation as the 2.6-kb hAR segment.

Filters were washed with solution containing 0.1% SDS, 0.05% sodium pyrophosphate and 0.4 x SSC at 37 C. Positive clones were plaque-purified and probed with 5'-end ^{32}P-labeled 24 base oligonucleotides that were specific to various steroid receptors. After eliminating clones for other SRs, the DNA inserts in the remaining clones were analyzed by restriction mapping and subcloned into M13 vectors for dideoxy sequence analysis.

If AR also has a cysteine-rich DNA-binding domain that is highly homologous with the DNA-binding regions of other SRs, some of the positive clones should contain cDNA for AR. To eliminate other SR-cDNA clones, we screened these positive clones with specific 24-base probes that had nucleotide sequences identical to nucleotide segments immediately next to the 5' end or the 3' end of the DNA-binding region of cDNA for these SR.

To obtain additional information on the remaining clones, DNA fragments in these clones were used as probes to screen a human X-chromosome library. By this method, cDNA clones were separated into X-chromosome-linked and unrelated groups. Since the X-chromosome has been suggested to contain an AR gene, the overlapping X-chromosome linked DNA fragments were ligated, selected, and amplified by using pBR322 and pGEM-3Z vectors (Figs. 1 and 2).

TRANSCRIPTION AND TRANSLATION

Plasmid DNA encoding AR were linearized with restriction enzymes and transcribed with T7 RNA polymerase. mRNAs were isolated and then translated in a rabbit reticulocyte lysate system. When the protein products were analyzed by SDS-PAGE, we found that the 2.6 kb hAR and the 2.72 kb rAR cDNAs yielded a major translated product that migrated as a 76 kDa protein, whereas the 3.6 kb hAR and 2.8 kb rAR cDNA produced not only a 94 kDa protein but also 76 kDa, 70 kDa, 55 kDa, 46 kDa, 32 kDa and 30 kDa proteins. All of these products were immunoprecipitable by a serum containing autoimmune antibodies to AR (Liao and Witte, 1985) and were able to bind DNA, indicating that each of them had a DNA-binding domain. The production of the 94 kDa, 76 kDa, 70 kDa, and 46 kDa can be explained as products of translation initiation at the first and other internal ATG codons. The 55 kDa, 32 kDa, and 30 kDa peptides may be proteolytic products of large forms.

ANDROGEN-BINDING AND SEDIMENTATION PROPERTIES

The steroid binding activity of proteins encoded by hAR and rAR cDNAs were studied by incubation of AR synthesized in a reticulocyte lysate. Receptor binding of a synthetic radioactive androgen, [3H]R1881(methyltrienolone), was studied by a hy-

droxylapatite-filter assay method (Liao et al., 1984) and
Scatchard plot analysis. We estimated that the apparent dissociation constants for hAR and rAR as 0.3 nM and 0.5 nM respectively. 5α-dihydrotestosterone (DHT), a potent intracellular
androgen, as well as another potent synthetic androgen, 7α,17α-
dimethyl-19-nortestosterone competed very well, whereas 17ß-
estradiol, progesterone, and dexamethasone were not effective
competitors. Testosterone was also much less effective than DHT
or R1881.

In a low salt medium, various steroid-receptor complexes
sediment as 3-5 S, 7-10 S, or even heavier forms. The 7-10 S or
larger forms can dissociate into the 3-5 S form in a medium containing 0.4 M KCl. In a medium containing 0.4 M KCl, ^{35}S-AR
made by hAR cDNA transcription and translation sediments as a
4 S form. Omission of KCl from the medium, or addition of a cytosolic fraction of rat ventral prostate to the ^{35}S-AR preparation
did not alter the sedimentation pattern of the radioactive AR,
suggesting that the 7-10 S or larger form of AR may be
formed only after a modification of the newly made hAR and or
association with a specific macromolecule. [^{3}H]R1881-bound
hAR also sediments at 4 S in a high salt medium. In the presence
of human serum containing autoantibodies to AR, the radioactive androgen-receptor complex sedimented at 8-10 S, indicating that the newly synthesized AR could be recognized by the
autoantibodies.

OVERALL STRUCTURES

Fig. 3 shows the cDNA sequences and deduced amino acid
sequences of rAR and hAR (Chang et al., 1988b). The open
reading frames of rAR and hAR cDNAs, from the first ATG (at
human nucleotide 532 or rat nucleotide 33) to the terminator
TGA, encodes 918 and 902 amino acids respectively, for hAR and
rAR with a molecular mass (M_r) of about 94,000. The first ATG is
preceded by a terminator TAA, indicating that it may be a
natural initiator. The open reading frames from the second ATG
of rAR or hAR encode 734 and 733 amino acids (M_r, about
76,000), respectively for hAR and rAR. Since the nucleotide
sequences surrounding these two ATG positions in AR cDNA
match Kozak's consensus sequence for an active start codon
(ANNATGN), both the 76 kDa and the 94 kDa forms of AR may be
produced from a single gene in target organs.

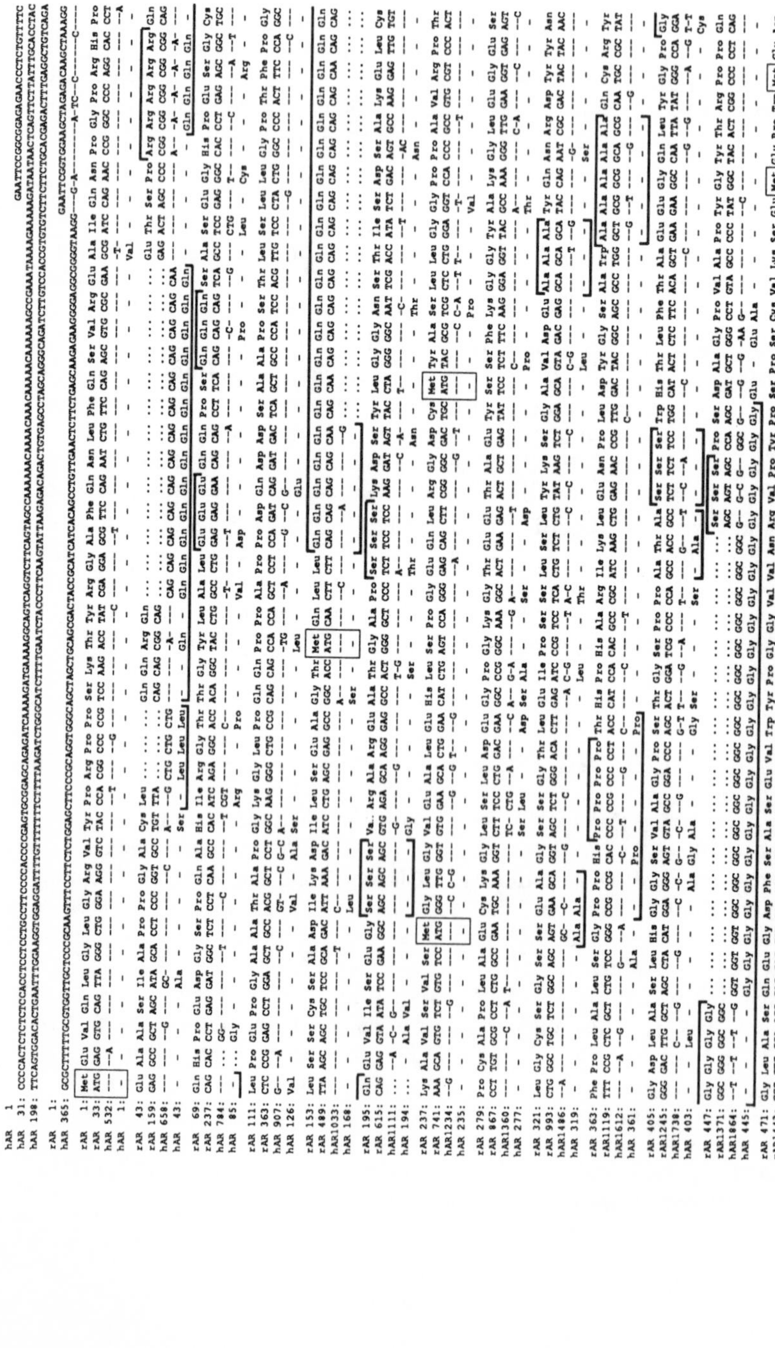

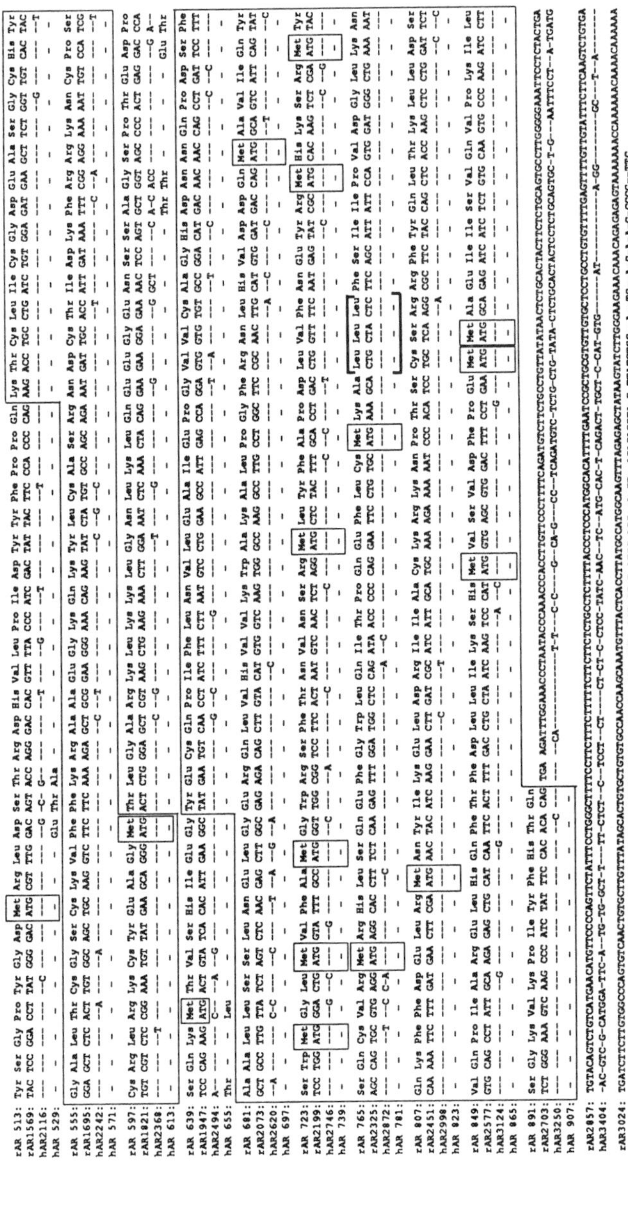

Fig. 3. cDNA and deduced amino acid sequences of rAR and hAR. The amino acid sequences in the putative DNA- (No.556-623 for hAR, No.540--607 for rAR) domains and in androgen-binding (N.666-918 for hAR, No.650-902 for rAR) domains are boxed. The missing residues in the gaps are shown by dots. The identical residues are shown by hyphens. Oligo- and polyamino acids regions are shown by boldface brackets.

DNA- AND ANDROGEN-BINDING DOMAINS

Like other steroid receptors, the primary structures of ARs deduced from AR cDNAs show that there is a cysteine-rich domain in hAR and rAR (Fig. 3). The two ARs have an identical amino acid sequence in this domain but there are 14 nucleotide changes. The 68 amino acid region may fold into two "zinc coordinated finger" structures and bind to hormone responsive elements on DNA as suggested for other steroid receptors and some transcriptional factors. All 10 cysteines of rAR, hAR, hPR and hGR as well as 9 of 10 cysteines in hER are conserved. When the DNA-binding domain of various SR are compared, a high homology (76-79%) is found between AR and hPR, hGR, or hMR. The homology between AR and hER, hRAR (retinoic acid receptor), chicken(c) VDR, and hTR or the oncogene product of avian erythroblastosis v-erbA virus at this domain was within 40-56%.

The sequences of 253 amino acids at the C-terminal ends of hAR and rAR are identical although there are 54 nucleotide changes (Fig. 3). This region corresponds to the putative hormone-binding domain in other SRs. In this region the percent homology between AR and hPR, hGR, or hMR is 50-54%, whereas the homology between AR and hER, hTR, or hRAR is 15% or less. A unique feature of the steroid-binding domain in all steroid receptors is its high methionine content. For ARs, there are 13 methionines in this domain. Their functional significance is not clear.

The amino acid sequence within the steroid-binding domain of rAR or hAR contains 4 highly conserved regions at the positions (based on rAR) between 698 to 736 (region I), 745 to 758 (region II), 786 to 805 (region III), and 838 to 847 (region IV). The percent homology in these regions among human AR, PR, GR, and MR is 65-100%. The homology between AR and ER at these regions is only 20-33%.

The least conserved region in the C-terminal half is located between DNA- and steroid-binding domains. Between hAR and rAR, 6 of the 41 amino acids in this region are different. Like other SRs this hinge region is hydrophilic and is rich in both the basic and acidic residues, suggesting that this site may be involved in an ionic interaction between a receptor and a modulator.

AMINO TERMINAL DOMAINS OF STEROID RECEPTORS

In general, the N-terminal regions outside the DNA-binding domain in different SRs do not appear to have a uniform structure. In fact, the differences in M_r among different SRs are largely due to the differences in the peptide length at these regions. Although there is a considerable homology between the same SRs in different species, the percent homology in this region is not as high as that in DNA and steroid-binding domains. For example, cER and hER have only 56% homology in the 140 amino acid region just outside the DNA-binding domain. rPR and hPR have about 80% homology at the N-terminal whereas their progestin- and DNA-binding domains have 99-100% homology.

Homologous structures in SRs, such as the DNA- and hormone-binding domains are apparently necessary for SRs to perform their functions through common mechanisms. However, nonhomologous structures in the amino terminal domain may be important for SRs to carry out receptor- or species-specific functions and/or involved in differential modulation of receptor activity by other factors.

POLYAMINO ACID MOTIFS.

In the N-terminal region, 116 of 555 amino acids of hAR are not identical with the corresponding sequences in rAR (Fig. 3). The most striking feature of AR cDNA, however, is the presence of different repetitive nucleotide sequences that encode oligo- or polyamino acid sequences in the N-terminal regions. Thus, hAR has a stretch of 27 glycine residues and a stretch of 17 glutamine residues whereas rAR has a stretch of 22 glutamine residues. In addition, hAR also has an octamer of proline, a tetramer of leucine, a hexamer and a pentamer of glutamine, whereas rAR cDNA has a pentamer of arginine, a pentamer of glycine, a trimer of glutamic acid, a trimer of glutamine, and a tetramer of proline. Both AR cDNAs also have sequences that encode trimers or pentamers of alanine and serine. A trimer of leucine was present in the androgen-binding domain of hAR and rAR, otherwise, no other oligomeric amino acid is present in the DNA- or hormone-binding domains of ARs.

Many regulatory gene products are known to contain poly-amino acid sequences. Among the most interesting examples are the "pen" or GGN repeat which encodes polyglycine and the "opa" (or "M") or CAG repeat which encodes polyglutamine in several homeotic genes. A glutamine stretch was also found in rGR previously but, to our knowledge, no other SR cDNA that has been analyzed so far has a nucleotide sequence that encodes a long stretch of a polyamino acid. The role of polyamino acid sequences in homeotic genes and SRs is not clear. It is conceivable that they may participate in the interaction of receptor proteins with components of transcription machinery or with modulatory factors. The presence of these protein motifs in some SRs supports the suggestion that many genes with different roles in the control of development may have evolved from the same ancestral gene and/or were reassembled from components by a process involving transposition of repetitive sequences, deletions, or chromosomal rearrangements.

NOTES

During the screeining of human testis λGT11 library, we isolated a cDNA having nucleotide and amino acid sequences that are virtually identical to that of published cDNA for human mineralocorticoid receptor (Arriza et al., 1987) with the following three exceptions: a change from (A) to (G) at bp No. 538 which alters the amino acid sequence from isoleucine to valine, a change from (C) to (T) at the bp No. 722 which alters the amino acid sequence from alanine to valine, and a nucleotide sequence change from (C) to (T) at the bp No.1497 which does not alter the deduced amino acid sequence.

Other investigators (Lubahn et al.,1988; Trapman et al.,1988) have isolated cDNAs for hAR that appear to encode the C-terminal portion (about 41 kDa) of the hAR described by us.

ACKNOWLEDGMENTS

This work was supported by Grants DK-37694 from the National Institutes of Health.

REFERENCES

Arriza JL, Weinberger C, Cerelli G, Glasser TM, Handelin BL, Housman DE, Evans RM (1987) Science 237: 268-275.

Chang C, Kokontis J, Liao S (1988a) Science 240: 324-326.

Chang C, Kokontis J, Liao S (1988b) Proc Natl Acad Sci USA 85:7211-7215.

Liao S, Witte D (1985) Proc Natl Acad Sci USA 82: 8345-8348.

Liao S, Witte D, Schilling K, Chang C (1984) J Steroid Biochem 20:11-17.

Lubahn DB, Joseph DR, Sullivan PM, Willard HF, French FS, Wilson EM (1988) Science 240:327-330.

Trapman J, Klaassen P, Kuiper GGJM, van der Korput JAGM, Faber PW, van Rooij HCJ, van Kessel AG, Voorhorst MM, Mulder E, Brinkmann AO (1988) Biochem Biophys Res Commun 153: 241-248.

Molecular Endocrinology and Steroid
Hormone Action, pages 65–80
© 1990 Alan R. Liss, Inc.

STRUCTURE, FUNCTION AND REGULATION OF THE GLUCOCORTICOID
RECEPTOR

Jan-Åke Gustafsson, Jan Carlstedt-Duke, Per-Erik
Strömstedt, Ann-Charlotte Wikström, Marc Denis,
Sam Okret and Yu Dong
Department of Medical Nutrition, Karolinska
Institute, Huddinge University Hospital F69,
S-141 86 Huddinge, Sweden

INTRODUCTION & BACKGROUND

Glucocorticoids are a group of steroid hormones with a
role of major importance in the regulation of energy
metabolism, the immune response as well as the inflammatory
response. Because of their major role in these fundamental
physiological functions, glucocorticoids have a very
wide-spread and vital usage in clinical therapy. Although a
principal model for the mechanism of action of glucocorticoids has been presented, detailed knowledge is still
lacking. Since glucocorticoids have such wide-spread basic
effects, a precise knowledge of their mechanism of action
at the molecular level is of both biological and clinical
significance.

Glucocorticoids exert their biological effects in
various target tissues with the help of an intracellular
soluble receptor protein, the glucocorticoid receptor (GR).
In normal mammalian tissues, all cells studied so far have
been found to contain GR to a varying degree. In fact,
glucocorticoids are essential for life in higher organisms
and a total lack of glucocorticoids leads to death
ultimately.

In similarity to the other steroid receptors, GR acts
following binding of ligand. Similar receptor proteins have
also been described for vitamin D (McDonell et al., 1987),
thyroid hormone (Sap et al., 1986; Weinberger et al., 1986)
and retinoic acid (Petkovich et al., 1987; Giguère et al.,
1987), although it is unclear if these receptor proteins

function in a manner similar to the steroid receptors. The classical two-step model first entails binding of the hormone to the receptor protein which enables the complex to bind specifically to DNA. For the estrogen and progestin receptors, the whole process has been postulated to occur within the nucleus (King and Greene, 1984; Welshons et al., 1984; Gasc et al., 1984; Perrott-Applanat et al., 1985). However, immunocytochemistry and immunohistochemistry indicate that the non-liganded form of GR resides in the cytoplasm and that it translocates to the nucleus in the presence of hormone (Gustafsson et al., 1981; Bernard and Joh, 1984; Antakly and Eisen, 1984; Fuxe et al., 1985; Wikström et al., 1987).

With regard to the second stage of this model, purified GR has been shown to interact with specific DNA sequences. This was originally demonstrated using sequences from mammary tumor virus (MTV), a retrovirus the expression of which is normally induced by glucocorticoids (Payvar et al., 1981, 1982, 1983; Scheidereit et al., 1983; Scheidereit and Beato, 1984). These binding sites coincide with sequences that are necessary and sufficient for glucocorticoid regulation in vivo (GRE). Furthermore, these binding sites for GR can confer glucocorticoid inducibility on heterologous promoters (Lee et al., 1981; Chandler et al., 1983; Hynes et al., 1983; Kühnel et al., 1986). Similar GREs have been described for several other gluco-corticoid-regulated genes, usually within the 5'-flanking region of the promoter but also 3' of the promoter, although at greatly varying distance from the promoter. Analysis of these binding sequences has defined a consensus GRE sequence, G G T A C A N N N T G T T/C C T (Jantzen et al., 1987; Beato et al., 1987). Usually GREs occur in pairs, or even greater number. The mechanism by which GR binding to GREs induces transcriptional activation is unknown. Recently, cooperative DNA-binding between GR and another transcription factor has been suggested (Schüle et al., 1988).

Even before the receptor was cloned, analysis of the protein showed that it consisted of three functional domains (Carlstedt-Duke et al., 1977, 1982; Wrange and Gustafsson, 1978; Dellweg et al., 1982; Wrange et al., 1984). The steroid-binding domain lies within the C-terminal third of GR and is defined by a unique trypsin cleavage site (Giguère et al., 1986; Godowski et al.,

1987). The DNA-binding domain is adjacent. Binding of steroid at one domain affects the specific interaction of DNA at the adjacent domain, presumably by an allosteric effect. However, each of the domains is self-sufficient with regard to the specificity of interaction with its particular ligand. This was clearly demonstrated using heterologous receptor constructions (Petkovich et al., 1987; Giguère et al., 1987; Green and Chambon, 1987). Thus, the DNA-binding domain of GR in heterologous receptors, will only recognise GREs and mediate effects only on genes normally regulated by glucocorticoids, independently of the specificity of the hormone-binding domain. Truncation of the receptor from the C-terminus, with deletion of the majority of the steroid-binding domain, results in a GR fragment with constitutive activity with regard to transcriptional activation (Godowski et al., 1987; Miesfeld et al., 1987; Hollenberg et al., 1987). Indeed, a small part of the GR corresponding to the DNA-binding domain, is sufficient to give transcriptional activation, albeit at greatly reduced efficiency (Godowski et al., 1987; Miesfeld et al., 1987). Deletion of all or a part of the N-terminal half of GR results in normal hormone-dependent transcriptional activation but at a greatly reduced level (32,36). In analogy to GAL 4 (Giniger and Ptashne, 1987), this region of GR necessary for maximal transcriptional activity is an acidic region and may be responsible for interaction with other transcription factors (Schüle et al., 1988).

Non-activated GR is associated with an M_r 90,000 protein (Housley et al., 1985; Mendel et al., 1986). All steroid receptors can associate with one and the same M_r 90,000 species (Joab et al., 1984), which is most likely identical to a heat shock protein with M_r 90,000 (hsp 90) (Sanchez et a., 1985; Catelli et al., 1985). Following heat-activation, GR does not associate with hsp 90 (Leach et al., 1979; Sherman and Stevens, 1984). Furthermore, the addition of molybdate to non-activated GR prevents heat-activation and it has been suggested that the activation process parallels the dissociation of the M_r 90,000 protein from GR (Sherman and Stevens, 1984; Vedeckis, 1983).

Recently, we have obtained evidence that dissociation of the GR-hsp 90 complex indeed occurs upon heat activation. Furthermore, binding of ligand is a necessary prerequisite for the heat-induced dissociation of GR from hsp 90 and subsequent binding of GR to both specific and non-specific DNA (Denis et al., 1988).

The hsp 90 or proteins related to hsp 90 have been cloned in several species including mice and man (Hickey et al., 1986; Rebbe et al., 1987; Moore et al., 1987). In the higher species, two different cDNAs corresponding to two slightly different proteins have been obtained. So far, cloning of rat hsp 90 has not been reported.

The role of post-translational modifications in the function of GR has not been elucidated. There is some evidence that GR is a glycoprotein since the lectins from Ulex europeus (UEA) and Lens culinaris (LEA) induced a concentration-dependent decrease in ^{3}H-dexamethasone binding to rat liver cytosol (Blanchardie et al., 1986) and Con A was shown to interact with microsomal GR in intact but not adrenalectomized rats (Smith et al., 1986). In addition to the recent reports on the existence of peptide sequences in GR that correspond to basic sequences known to act as nuclear translocation signals (Goldfarb et al., 1986; Wolff et al., 1987; Picard and Yamamoto, 1987), there are also reports on the existence of O-linked N-acetyl-glucosamine residues that occur in high abundance on proteins that are preferentially localized in the nucleus or at the nuclear envelope (Holt and Hart, 1986). Furthermore, it has been shown that a heparan sulfate proteoglycan species enriched in 2-O-sulfated glucosamine residues is preferentially translocated to the nucleus. Whether the specific carbohydrate structures on these proteins participate in the nuclear translocation process remains to be elucidated.

A number of studies have implicated the limiting nature of steroid receptor numbers in eliciting the biological effects. Therefore, the cell may control the magnitude of the response by regulating receptor expression. Autoregulation of steroid receptors has been indicated by several investigators using ligand-binding assays. This may be a way for the cell to adapt to changes in the autologous hormone concentration and thereby biological response. The presence of glucocorticoids has been reported to cause a down-regulation of cellular GR levels (Svec and Rudis, 1981; Cidlowski and Cidlowski, 1981; Sapolsky et al., 1984). Adrenalectomy causes an

~2-fold up-regulation of GR concentration (50,58). Autoregulation of GR has been demonstrated both in several cell
lines and in several tissues (for review see Svec, 1985),
although some tissue differences may exist (Sapolsky et
al., 1984; Turner, 1986). Similarly, estrogens have been
shown to down-regulate estrogen receptor protein (Eckert,
1984) or mRNA levels (Maxwell et al., 1987), and progestins
down-regulate progesterone receptor or mRNA levels (Weil et
al., 1988). In contrast, 1,25-dihydroxyvitamin D_3 and
androgens caused an up-regulation of 1,25-dihydroxyvitamin
D_3 and androgen receptors, respectively (McDonnell et al.,
1987; Costa et al., 1985; Syms et al., 1985).

The glucocorticoid induced down-regulation of GR is
probably a primary response mediated by GR, since it also
occurs in the presence of a protein synthesis inhibitor,
cycloheximide (Svec and Ruidis, 1981; McIntyre and Samuels,
1985; Okret et al., 1986) in response to glucocorticoid
agonists (Cidlowski and Cidlowski, 1981; Svec, 1985).
Down-regulation is relatively slow, requiring at least 24 h
of exposure to glucocorticoids to reach a maximal 50-80%
response (Svec and Rudis, 1981; Cidlowski and Cidlowski,
1981; Danielsen and Stallcup, 1984). Upon removal of
glucocorticoids, protein-synthesis dependent replenishment
of GR slowly occurs (Svec and Rudis, 1981).

Until recently, all studies on GR regulation were
based on ligand binding assays. It cannot be excluded that
observed changes in receptor amount are due to factors
altering hormone binding capacity, thereby affecting
receptor quantitation. In line with this, it is known that
ligand binding to GR is altered by several chemical
modifiers and that the receptor can exist in two states
regarding its ability to bind ligand (for a review see
Housley et al., 1984). We have therefore chosen to address
the issue of GR autoregulation by studying the mechanism of
both GR mRNA and protein expression and turnover.

Regulation of a functional GR protein in the cell may
occur at various levels; transcriptional or posttranscriptional, the latter involving e.g. nucleocytoplasmic transport, changes in GR mRNA stability,
translation efficiency, protein turnover and the conversion
of the GR from a nonfunctional to a functional state (see
also above). In fact, using dense amino acid labelling and

ligand binding techniques it has been shown that glucocorticoid hormones decrease the half-life of GR from 19 h
to 9.5 h without affecting the rate of GR synthesis
(McIntyre and Samuels, 1985). Furthermore, it has been
shown that estrogens prolong the half-life of the vitellogenin mRNA (Brock and Shapiro, 1983) and glucocorticoids
prolong the half-life of the human growth hormone mRNA
(Diamond and Goodman, 1985; Paek and Axel, 1987) and that
of class I alcohol dehydrogenase mRNA (Dong et al., 1988).

Studies by ourselves (Okret et al., 1986) have shown
that the glucocorticoid-induced down-regulation of GR mRNA
also occurs in the absence of protein synthesis. This is
suggestive of a primary response involving the GR protein
itself. An increase in GR mRNA is observed in the presence
of cycloheximide (Okret et al., 1986) which might indicate
the presence of a gene repressive protein or a protein
affecting mRNA stability. In analogy with the postulated
model for positively regulated genes, experiments have
indicated the importance of GR in controlling also
negatively regulated genes. Both the pro-opiomelanocortin
gene (Eberwine and Roberts, 1984) and the prolactin gene
(Camper et al., 1985) are negatively regulated by glucocorticoid hormones at the transcriptional level.
Preliminary run-on data by ourselves and Rosewicz et al.
(Rosewicz et al., 1988) suggest that down-regulation of GR
expression involves, at least in part, decreased GR gene
transcription. In contrast, down-regulation of e.g.
urokinase-type plasminogen activator (Busso et al., 1986)
and rat type I procollagen (Ragshow et al., 1986) by
glucocorticoid hormones seems to be at least in part due to
a posttranscriptional mechanism.

Our primary interest has concentrated on the mechanism
of action of glucocorticoids. As a basis for our investigation we studied the interaction of natural glucocorticoids
and their metabolites with different protein fractions in
the rat liver (Carlstedt-Duke et al., 1975, 1977). This led
to a characterization of the rat liver GR and definition of
functional domains, based on limited proteolysis
(Carlstedt-Duke et al., 1977, 1979, 1982; Wrange and
Gustafsson, 1978), purification of the rat liver GR (Wrange
et al., 1979), its characterization (Wrange et al, 1984),
and the raising of antibodies (Okret et al., 1981, 1984).
The purified receptor has been used to study the specific
interaction with DNA (Payvar et al., 1981, 1982, 1983;

Wrange et al., 1986) as well as for definition of the functional domains at the protein sequence level (Carlstedt-Duke et al., 1987) and identification of steroid-binding amino acid residues (Carlstedt-Duke et al., 1988). The antibodies raised have been used to clone GR cDNA (Miesfeld et al., 1984, 1986), for immunoaffinity chromatography (Denis et al., 1987, 1988) and for studies of the intracellular localization of GR (Gustafsson et al., 1981; Fuxe et al., 1985; Wikström et al., 1987). The availability of GR cDNA has permitted initial studies on the regulation of GR expression (Okret et al., 1986).

RECENT STUDIES

Domain structure of GR

Limited proteolysis of GR with trypsin or chymotrypsin results in discrete fragments of GR with DNA-binding and/or steroid-binding properties. Digestion of crude or purified preparations of GR gives rise to identical fragments of GR and this led to the definition of three functional domains of the proteins, namely 1. a steroid-binding domain, 2. a DNA-binding domain and 3. a third domain of unclear function that was immunodominant (Carlstedt-Duke et al., 1982).

Following limited proteolysis of purified GR and sequence analysis of the products obtained, the functional domains could be identified at the protein level (Carlstedt-Duke et al., 1987). The steroid-binding domain is defined by a unique tryptic cleavage site after Lys-517 (rat GR) and encompasses the C-terminal third of GR, residues 518-795 (rat GR). The DNA-binding domain was found to be immediately adjacent. It is defined by chymotrypsin cleavage that was found to occur at multiple sites. Two sites were first tentatively identified with cleavage after Phe-409 and Tyr-413. These two cleavage sites have subsequently been conclusively identified by repeated sequence analysis of the chymotryptic cleavage product of either purified GR or recombinant GR fragments. The definition of the functional domains at the protein level is in accordance with the definition of these domains by deletional or insertional mutation analysis (Giguère et al., 1986; Godowski et al., 1987; Hollenberg et al., 1987; Danielsen et al., 1986).

Steroid-binding site

The steroid-binding domain of GR makes up a large part of the protein, ~30 kDa (cf. above). Mutation almost anywhere within this domain disrupts steroid-binding (Giguère et al., 1986; Godowski et al., 1987; Miesfeld et al., 1987; Hollenberg et al., 1987; Danielsen et al., 1986). Thus, it has not been possible using recombinant techniques to dissect the structure of this domain in more detail.

For this purpose, we have radiosequence analysed affinity-labelled preparations of purified GR (Carlstedt-Duke et al., 1988). The purified GR was either photoaffinity-labelled through the 3-oxo-Δ^4-structure of ^{3}H-triamcinolone acetonide (TA) or affinity-labelled through the side chain of ^{3}H-dexamethasone 21-mesylate (DM). The affinity-labelled GR was specifically cleaved, using trypsin (with or without prior succinylation), chymotrypsin or cyanogen bromide, after prior denaturation. The resulting mixture of peptides was analysed by radiosequence analysis. From the radio-activity detected in each cycle compared with the specificity of cleavage as well as the primary structure of rat GR derived from the cDNA (Miesfeld et al., 1986), several amino acids bound to the steroid could be identified. Affinity-labelling with TA gave two major peaks which were found to correspond to Met-622 and Cys-754. Affinity-labelling with DM gave rise to one single peak corresponding to Cys-656, in agreement with Simons et al. (Simons et al, 1987). Each of these three steroid-binding amino acid residues was found to lie within hydrophobic segments of the steroid-binding domain. Thus, the steroid-binding domain undergoes folding to give a tertiary structure with a hydrophobic pocket forming the steroid-binding site. Within this pocket, Met-622 and Cys-754 occur in close proximity to each other and at about 1.5 nm distance from Cys-656.

Similar analysis of progestin receptor (PR) purified from T47D cells is in progress. There is marked cross-reactivity regarding the specificity of hormone-binding to the two receptors. Analysis of the primary structure reveals a 49% positional identity within the steroid-binding domain of PR and GR, whereas the hydropathy pattern is remarkably similar (Carlstedt-Duke et al., 1988).

Photoaffinity-labelling of PR through the 3-oxo-Δ^4-structure of ^{3}H-R5020 results in several peaks of radio-activity when analysed by radiosequence analysis as described above. The two major peaks have been preliminarily identified as Met-759 (equivalent to Met-622 in rat GR) and Met-909 (18 residues C-terminal of the equivalent of Cys-754 in rat GR).

DNA-binding domain

In order to obtain sufficient material for detailed analysis of protein structure, the DNA-binding domain of human GR has been expressed in E. coli. The construction used enables the recovery of the functional DNA-binding domain defined above, in a native, functional form. This recombinant GR material has been found to bind to DNA with the same specificity as purified intact rat GR, although with about ten-fold lower affinity. The protein is expressed as a fusion protein at the C-terminal end of protein A, using a temperature-induced λ-promoter. The fusion protein is found in the soluble fraction after lysis of the bacteria and is recovered by binding to IgG-Sepharose. The GR DNA-binding domain is released by cleavage of the fusion protein with chymotrypsin, while bound to IgG-Sepharose. The recombinant DNA-binding domain (DBD_r) is then separated from chymotrypsin by DNA-cellulose chromatography. Using this procedure, 2-3 mg of highly purified DBD_r is obtained per litre culture, devoid of any contaminating material. Analysis of the product reveals a single protein species with a total amino acid composition, an N-terminal sequence and a C-terminal sequence corresponding to the expected primary structure. Chymotrypsin cleavage was found to occur exclusively after Tyr-393 (equivalent to Tyr-413 in rat GR), under the conditions used in this purification scheme (Dahlman K., Strömstedt P., Rae C., Jörnvall H., Carlstedt-Duke J. and Gustafsson J.-Å., submitted to J. Biol. Chem. for publication).

Using both purified rat GR and the DBD_r preparation described above, analysis of the structure of the DNA-binding domain is being performed. Analysis of DBD_r shows that the great majority of the Cys residues occur as free thiol groups and that carboxymethylation destroys the specific DNA-binding properties.

hsp 90-GR interaction

A rapid, high-yield purification method for hsp 90 has been developed in our laboratory. Attempts to reconstitute the 9S GR-hsp 90 complex in vitro by mixing purified hsp 90 and GR purified by immunoaffinity chromatography have not been successful. However, preliminary experiments in which GR is expressed in vitro in a rabbit reticulocyte lysate system have yielded GR in the 9S form. Furthermore, it has been possible to demonstrate the presence of hsp 90 in the lysate, by probing the lysate in protein immunoblotting experiments with antibodies specific for rat hsp 90. This could indicate that association of GR with hsp 90 might occur prior to final folding of GR. Furthermore, these results imply that association of hsp 90 and GR occurs in a rather nonspecific environment and that, consequently, denaturation of GR and/or hsp 90 followed by renaturation might also enable in vitro reconstitution of the GR-hsp 90 complex.

Regulation of GR expression.

Studies in rat hepatoma cells during continuous treatment with dexamethasone showed a cyclic pattern of both GR mRNA and protein as assessed by RNA and Western blot analysis. The GR mRNA showed a significant 1.5-2-fold increase in GR mRNA after 6 h of treatment followed by a 50-80% reduction after 24-28 h as compared to the initial level (Okret et al., 1986). Parallel changes in the GR protein level were seen, albeit with a 6-24 h delay. The same phenomena, however, without the initial increase in GR mRNA and protein, were found to occur in liver of rats treated with a single dose of dexamethasone. Preliminary data suggest that down-regulation of GR mRNA is controlled at least in part at the transcriptional level. However, in addition to the transcriptional regulation of GR gene expression, a glucocorticoid dependent post-translational modification in the rate of GR protein turnover was observed. GR protein half life was ~25 h in the absence of dexamethasone while it decreased to ~11 h in the presence of hormone. Interestingly, the effect on GR protein down-regulation inferred from the transcriptional and posttranslational control mechanisms was less than expected. This may suggest that other control mechanisms, e.g. translational mechanisms, "neutralizing" the GR mRNA changes may exist. In fact, others (McIntyre and Samuels,

1985) have suggested, based on dense amino acid labelling
of GR in GC pituitary cells, that down-regulation of GR by
glucocorticoid hormones is a post-translational mechanism
not affecting GR synthesis rate.

We have already shown that the GR protein interacts
specifically with regions in the GR cDNA which correspond
to the 3' untranslated region (3' UTR, ref. Okret et al.,
1986). The 3' UTR exists in the genome as a single exon
(Miesfeld et al., 1985). It is tempting to speculate that
the regulation of GR mRNA by glucocorticoid hormones may
involve binding of the GR to a GRE within the 3' UTR. As
mentioned above, regulatory sequences are not only
localized within the 5' promoter region of GR-regulated
genes. Furthermore, a function for the 3' UTR region may be
inferred from computer based comparison between hGR and rGR
nucleotide sequences demonstrating that the 3' UTR, in
addition to the coding sequence, is highly conserved
(Miesfeld et al., 1986). In addition, preliminary data of
GR gene organisation by Southern blotting of a glucocorticoid resistant cell line has indicated reorganization
of the 3'UTR (Norris et al., 1987). Although the 3' UTR
probably is located tens of kb away from the GR gene
promoter (Miesfeld et al., 1985), chromosomal arrangement
might place it in close proximity to other important gene
regulatory regions (Ptashne, 1986). The 3' UTR may also be
involved in posttranscriptional regulation, determining
mRNA stability (Sham and Kamen, 1986; Rahmsdorf et al.,
1987) or translational efficiency (Liebhaber and Kan, 1982;
Miller et al., 1984).

REFERENCES

Antakly T, Eisen HJ (1984). Endocrinology 115: 1984.
Bernard PA, Joh TH (1984). Arch Biochem Biophys 229: 466.
Beato M, Arnemann J, Chalepakis G, Slater E, Willmann T
    (1987). J Steroid Biochem 27: 9.
Blanchardie P, Lustenberger P, Denis M, Orsonneau JL,
    Bernard S (1986). J Steroid Biochem 24: 263.
Busso N, Belin D, Failly-Crépin C, Vassalli J-D (1986). J
    Biol Chem 261: 9309.
Brock ML, Shapiro DJ (1983). Cell 34: 207.
Camper SA, Yao YAS, Rottman FM (1985). J Biol Chem 260:
    1246.

Carlstedt-Duke J, Gustafsson J-Å, Gustafsson SA (1975).
Biochemistry 14: 639.
Carlstedt-Duke J, Gustafsson J-Å, Gustafsson SA, Wrange ö
(1977). Eur J Biochem 73: 231.
Carlstedt-Duke J, Gustafsson J-Å, Wrange ö (1977). Biochim
Biophys Acta 497: 507.
Carlstedt-Duke J, Wrange ö, Dahlberg E, Gustafsson J-Å,
Högberg B (1979). J Biol Chem 254: 1537.
Carlstedt-Duke J, Okret S, Wrange ö, Gustafsson J-Å (1982).
Proc Natl Acad Sci USA 79: 4260.
Carlstedt-Duke J, Strömstedt P-E, Wrange ö, Bergman T,
Gustafsson J-Å, Jörnvall H (1987). Proc Natl Acad Sci USA
84: 4437.
Carlstedt-Duke J, Strömstedt P-E, Persson B, Cederlund E,
Gustafsson J-Å, Jörnvall H (1988). J Biol Chem 263:
6842.
Catelli MG, Binart N, Jung-Testas I, Renoir J-M, Baulieu
E-E, Feramisco JR, Welsch WJ (1985). EMBO J 4: 3131.
Chandler VL, Maler BM, Yamamoto KR (1983). Cell 33: 489.
Cidlowski JA, Cidlowski NB (1981). Endocrinology 109:
1975.
Costa EM, Hirst MA, Feldman D (1985). Endocrinology 117:
2203.
Danielsen M, Northrop J, Ringold G (1986). EMBO J 5: 2513.
Danielsen M, Stallcup MR (1984). Mol Cell Biol 4: 449.
Dellweg H-G, Hotz A, Mugele K, Gehring U (1982). EMBO J 1:
285.
Denis M, Wikström A-C, Gustafsson J-Å (1987). J Biol Chem
262: 11803.
Denis M, Wikström A-C, Gustafsson J-Å (1988). J Steroid
Biochem 30: 33.
Denis M, Poellinger L, Wikström A-C, Gustafsson J-Å (1988).
Nature 333: 6174.
Diamond DJ, Goodman HM (1985). J Mol Biol 181: 41.
Dong Y, Poellinger L, Okret S, Höög J-O, von Bahr-Lindström
H, Jörnvall H, Gustafsson J-Å (1988). Proc Natl Acad Sci
USA 85: 767.
Eberwine JH, Roberts JL (1984). J Biol Chem 259, 2166.
Eckert RL, Mullick A, Rorke EA, Katzenellenbogen AS (1984).
Endocrinology 114: 629.
Fuxe K, Wikström A-C, Okret S, Agnati LF, Härfstrand A, Yu
Z-Y, Granholm L, Zoli M, Vale W, Gustafsson J-Å (1985).
Endocrinology 117: 1803.
Gasc J-M, Renoir J-M, Radanyi C, Joab I, Tuohimaa P,
Baulieu E-E (1984). J Cell Biol 99: 1193.

Giguère V, Hollenberg SM, Rosenfeld MG, Evans RM (1986). Cell 46: 645.

Giguère V, Ong ES, Segui P, Evans RM (1987). Nature 330: 624.

Giniger E, Ptashne M (1987). Nature 330: 670.

Godowski PH, Rusconi S, Miesfeld R, Yamamoto KR (1987). Nature 325: 365.

Goldfarb DS, Gariépy J, Schoolnik G, Kornberg RD (1986). Nature 322: 641.

Goldstein IJ, Poretz RD (1986) In Liener IE, Sharon N, Goldstein IJ (eds): "The Lectins. Properties, Functions and Application in Biology and Medicine," Orlando, Florida, USA: Academic Press.

Green S, Chambon P (1987). Nature 325: 75.

Gustafsson J-Å, Carlstedt-Duke J, Fuxe K, Carlström K, Okret S, Wrange ö (1981) In Fuxe K, Wetterberg L, Gustafsson J-Å (eds): "Steroid Hormone Regulation of the Brain," Oxford: Pergamon Press, pp 31.

Hickey E, Brandon SE, Sadis S, Smale G, Weber LA (1986). Gene 43: 147.

Hollenberg SM, Giguère V, Segui P, Evans RM (1987). Cell 49: 39.

Holt GD, Hart GW (1986). J Biol Chem 261: 8049.

Housley PR, Grippo JF, Dahmer MK, Pratt WB (1984). In "Biochemical Actions of Hormones," New York: Academic Press Inc, vol 2, pp 347.

Housley PR, Sanchez ER, Westphal HM, Beato M, Pratt WB (1985). J Biol Chem 260: 13810.

Hynes NE, von Ooyen IJ, Kennedy M, Herrlich P, Ponta H, Groner B (1983) Proc Natl Acad Sci USA 80: 3637.

Isohashi F, Tsukanaka K, Nakanishi Y, Fukushima H (1979). Cancer Res 39: 5132.

Jantzen H-M, Strähle U, Gloss B, Steward F, Schmid W, Boshard M, Miksicek R, Schütz G (1987). Cell 49: 29.

Joab I, Radanyi C, Renoir M, Buchou T, Catelli M-G, Binart N, Mester J, Baulieu E-E (1984). Nature 308: 850.

King WJ, Greene GL (1984). Nature 307: 745.

Kühnel B, Buetti E, Diggelmann h 81986). J Mol 'Biol 190: 367.

Lee F, Mulligan R, Berg P, Ringold G (1981). Nature 294: 228.

Leach KL, Dahmer MK, Hammond ND, Sandro JJ, Pratt WB (1979). J Biol Chem 254: 11884.

Liebhaber SA, Kan YW (1982). J Biol Chem 257: 11852.

Maxwell BL, McDonnell DP, Conneely OM, Schulz TZ, Greene GL, O'Malley BW (1987). Mol Endocrinol 1: 25.

McDonnell DP, Mangelsdorf DJ, Pike JW, Haussler MR, O'Malley BW (1987). Science 235: 1214.

McIntyre WR, Samuels HH (1985). J Biol Chem 260: 418.

Mendel DB, Bodwell JE, Gametchu B, Harrison RW, Munck A (1986). J Biol Chem 261: 3758.

Miesfeld R, Godowski PH, Maler BA, Yamamoto KR (1987). Science 236: 423.

Miesfeld R, Okret S, Wikström A-C, Wrange ö, Gustafsson J-Å, Yamamoto KR (1984). Nature 312: 779.

Miesfeld R, Rusconi S, Godowski PJ, Maler B.M., Okret S, Wikström A-C, Gustafsson J-Å, Yamamoto KR (1986). Cell 46, 389.

Miesfeld R, Rusconi S, Okret S, Wikström A-C, Gustafsson J-Å, Yamamoto KR (1985). In "Sequence specificity in Transcription and Translation. UCLA Symposia on Molecular and Cellular Biology, New Series," New York: Alan R Liss, vol 30, pp 535.

Miller AD, Curran T, Verma M (1984). Cell 36: 51.

Moore SK, Kozak C, Robinson EA, Ullrich SJ, Appella E (1987). Gene 56: 29.

Norris JS, Mclead SL, Cornett LE, Smith RG (1987) J Cell Biochem. Suppl 11A, Abstract B119, pp 109.

Nilsson B, Abrahamsén L, Uhlén M (1985). EMBO J 4: 1075.

Okret S, Poellinger L, Dong Y, Gustafsson J-Å (1986). Proc Natl Acad Sci USA 83: 5899.

Okret S, Carlstedt-Duke J, Wrange ö, Carlström K, Gustafsson J-Å (1981) Biochim Biophys Acta 677: 205.

Okret S, Wikström A-C, Wrange ö, Andersson B, Gustafsson J-Å (1984). Proc Natl. Acad Sci USA 81: 1609.

Paek I, Axel R (1987) Mol Cell Biol 4: 1496.

Payvar F, DeFranco D, Firestone GL, Edgar B, Wrange ö, Okret S, Gustafsson J-Å, Yamamoto KR (1983). Cell 35: 381.

Payvar F, Firestone GL, Ross SR, Chandler VL, Wrange ö, Carlstedt-Duke J, Gustafsson J-Å, Yamamoto KR (1982). J Cell Biochem 19: 241.

Payvar F, Wrange ö, Carlstedt-Duke J, Okret S, Gustafsson J-Å, Yamamoto KR 81981). Proc Natl Acad Sci USA 78: 6628.

Payvar F, Wrange ö (1983). In Eriksson H, Gustafsson J-Å (eds): "Steroid Hormone Receptors: Structure and Function," Amsterdam: Elsevier, pp 267.

Perrott-Applanat M, Logeat F, Groyer-Picard MT, Milgrom E (1985). Endocrinology 116: 1473.

Petkovich M, Brand NJ, Krust A, Chambon P (1987). Nature 330: 444.

Picard D, Yamamoto KR (1987). EMBO J 6: 3333.

Ptashne M (1986). Nature 322: 697.

Ragshow R, Gossage D, Kang AH (1986). J Biol Chem 261: 4677.

Rahmsdorf HJ, Schönthal A, Angel P, Liftin M, Herlich P (1987). Nucl Acid Res 15: 1643.

Rebbe NF, Ware J, Bertnia RM, Modrich P, Stafford DW (1987). Gene 53: 235.

Rosewicz S, McDonald AR, Maddux BA, Goldfine JD, Miesfeld RL, Logsdon CD (1988). J Biol Chem 263: 2581.

Sap J, Munoz A, Damm K, Goldberg Y, Ghysdael J, Leutz A, Beng H, Wennström B (1986). Nature 324: 635.

Sanchez ER, Toft DO, Schlesinger MJ, Pratt WB (1985). J Biol Chem 260: 12398.

Sapolsky RM, Krey LC, McEwen BS (1984). Endocrinology 114: 287.

Scheidereit C, Beato M (1984). Proc Natl Acad Sci USA 81: 3029.

Scheidereit C, Geisse S, Westphal HM, Beato M (1983). Nature 304: 749.

Schüle R, Muller M, Otsuka-Murakani H, Renkawitz R (1988). Nature 322: 87.

Scam G, Kamen R (1986). Cell 46: 659.

Sherman MR, Stevens J (1984). Ann Rev Physiol 46: 83.

Simons SS, Pumphrey JG, Rudikoff S, Eisen HJ (1987). J Biol Chem 262: 9676.

Smith K, Shuster S (1984). J Endocrinol 102: 161.

Svec F (1985). Life Sciences 36: 2359.

Svec F, Rudis M (1981). J Biol Chem 256: 5984.

Syms AJ, Norris JS, Panko WB, Smith RG (1985). J Biol Chem 260: 455.

Turner BB (1986). Endocrinology 118: 1211.

Vedeckis WV (1983). Biochemistry 22: 1983.

Weinberger C, Thompson CC, Ong ES, Lebo R, Gruol DJ, Evans RM (1986). Nature 324: 621.

Weil L, Krett NL, Francis MD, Gordon DF, Wood WM, O'Malley BW, Horwitz KB (1988). Molec Endocrin 2: 62.

Welshons WV, Leiberman ME, Gorski J (1984). Nature 307: 747.

Wikström A-C, Bakke O, Okret S, Brönnegård M, Gustafsson J-Å (1987). Endocrinology 120: 1232.

Wolff B, Dickson RB, Hanover JA (1987). Trends Pharmacol Sci 8: 119.

Wrange Ö, Carlstedt-Duke J, Gustafsson J-Å (1979). J Biol Chem 254: 9284.

Wrange ö, Carlstedt-Duke J, Gustafsson J-Å (1986). J Biol Chem 261: 11770.
Wrange ö, Gustafsson J-Å (1978) J Biol Chem 253: 856.
Wrange ö, Okret S, Radojcic M, Carlstedt-Duke J, Gustafsson J-Å (1984). J Biol Chem 259: 4534

II. INTERCONVERSION AND TRANSFORMATION OF STEROID HORMONE RECEPTORS

Molecular Endocrinology and Steroid
Hormone Action, pages 83–95
© 1990 Alan R. Liss, Inc.

ESTROGEN RECEPTOR INTERCONVERSION, FACTORS REGULATING CONFORMATION AND FUNCTIONS

Roy G. Smith[1,2], Abhijit Nag[1], Nooshine Dayani[1], Reid W. McNaught[1], and Shirish Shenolikar[3]

[1]Baylor College of Medicine, Houston, Texas, [2]Merck Sharp & Dohme Res. Labs., Rahway, NJ and [3]University of Texas Medical School, Houston, TX

INTRODUCTION

The mechanism of estrogen receptor action in the chick and hen oviduct has been our major focus. The hormonally responsive oviduct provides an excellent model for investigation of estrogenic modulation of ovalbumin gene expression (Harris et al., 1975). Two estrogen-specific receptor species have been identified as mediators of this process (Smith et al., 1979). The two receptor forms bind estradiol with high but differing affinity, and have different sedimentation coefficients. The receptors have been defined as R_x, which binds estradiol with an equilibrium dissociation constant (Kd) of 0.06 nM and sediments at 4.2 S, and R_y with a Kd of 0.8 nM which sediments at 3.5 S (McNaught and Smith, 1986). Interestingly, the difference in Kd is caused by the ten-fold difference in the rate of association of [^{3}H]estradiol. Each of these receptors appears to have discrete functions in eliciting an estrogenic response in oviduct epithelial tubular gland cells (Smith and Taylor, 1981).

Our objective has been to determine the molecular mechanism of receptor mediated activation of steroid responsive genes. Central to this issue is the knowledge that while steroid-receptor interactions may not be essential for DNA binding in vitro, transcription of the associated genes in vivo requires the presence of steroid (Sluyser, 1985; Theulaz et al., 1988). Since DNA binding of transcription factors is not in general the only factor involved in initiating transcription it follows that post–translational modi- fication(s) and perhaps the conformation of the trans- cription factor plays a key role in activation or repression of genes.

The steroid receptors are the best characterized transcription factors which act in a cell specific manner. Of the two forms of the estrogen receptors, R_x predominates in cells depleted of estrogen, while R_y is present in cells actively participating in estrogen induced gene transcription (Smith and Taylor, 1981; Taylor and Smith, 1985; McNaught et al., 1986). Most importantly, we have shown that R_x and R_y have the same monomeric subunit, which appears to exist in alternative conformations. Furthermore, R_x can be converted to R_y in the presence of estradiol/ATP/Mg^{2+} (McNaught and Smith, 1986; McNaught et al., 1986; Raymoure et al., 1986). An additional observation which provided the basis for our "Receptor Interconversion Model of Hormone Action" was that following estrogen treatment a third form of the receptor is present in chick oviduct cytosol. This form of the receptor does not bind estradiol until it is treated with ATP/Mg^{2+}, a process which converts it to R_y. This non–estradiol binding form has been defined as R_{nb} (Raymoure et al., 1986).

We have established that the estrogen receptor exists in three different conformational states. Each state can be stabilized by reactions requiring the gamma phosphoryl moiety of ATP (Raymoure et al., 1985; Raymoure et al., 1986). It is important to define the transcriptional role of each of these receptor forms, and to elucidate the mechanisms controlling their interconversion. We had already established by using isolated chick oviduct chromatin or nuclei that pure R_x was capable of directly increasing RNA polymerase II activity (Smith and Schwartz, 1979; Taylor and Smith, 1982). However, presumably because other transcription factors were lacking, we were unable to demonstrate initiation of ovalbumin gene transcription. More recently, we demonstrated that hydroxytamoxifen (TX-OH) has a 100-fold higher affinity for R_x than R_y and is capable of blocking the ATP/Mg^{2+} mediated R_x to R_y conversion (McNaught et al., 1986). It is also relevant that R_y appears to be a recyclable form of the estrogen receptor since the R_y to R_{nb} conversion is easily reversible in vitro (Raymoure et al., 1985) and R_{nb} is found in vivo shortly after estrogen treatment (Raymoure et al., 1986). In vivo following antiestrogen treatment receptors accumulate in the nucleus but are not "processed" (Sutherland et al., 1977). These observations argue that, since active growth and transcription are associated with the appearance of R_y and R_{nb} the interconversion $R_{nb} \rightleftarrows R_y$ plays an important role in estrogen action.

RESULTS OF EXPRESSION STUDIES

Our receptor interconversion model of hormone action suggests that the receptor exists in different conformations which are either inactive transcriptionally, or which are capable of acting as repressors and activators, with the ligand itself being directly involved in derepression or activation. Furthermore, it predicts that the transcriptional mode of the receptor is modified directly or indirectly by phosphorylation reactions within the cell (McNaught et al., 1986; Raymoure et al., 1986). Experiments were therefore designed to determine whether the transcriptional activity of the receptor could be modulated by altering intracellular phosphorylation reactions. Another focus was to determine how receptor conformation (R_x and R_y) in the presence of ligand related to the transcriptional activity of a reporter gene.

The 5'-flanking region of the vitellogenin gene was selected as a regulatable unit since vitellogenin gene transcription is regulated by estrogen, and the DNA sequence containing an estrogen-response-element (ERE) has been well defined (Klein-Hitpass et al., 1986). The chloramphenicol acetyl transferase (CAT) reporter gene was selected since it is readily assayable. A CAT construct containing the thymidine kinase promoter was selected, since it allows CAT to be constitutively active. This constitutive expression allows one to assay for repression or activation. Selection of a recipient cell for cotransfection of an expression vector containing the chicken estrogen receptor cDNA and the vitellogenin-CAT constructs, required a cell line which did not express estrogen receptors and in which inducible phosphorylation reactions were well defined. The A431 cell line provided this system. It does not contain estrogen receptors, it has high tyrosine kinase activity, and responds to EGF by activating the tyrosine kinase activity of the EGF receptor (Hunter et al., 1984). Furthermore A431 cells are responsive to the phorbol ester TPA which stimulates kinase C to counteract the tyrosine kinase activity. Kinase C activation results in increases in threonine and serine phosphorylation (Hunter et al., 1984; Lin et al., 1986).

The calcium phosphate protocol (Gorman, 1985) was used for transfection of A431 cells. The chicken estrogen receptor cDNA cloned into the pKCR2 expression vector and designated CERO was kindly provided by Dr. Pierre Chambon, and the pA2 (-821/-87) tk-CAT construct which contains the well-defined ERE (Klein-Hitpass et al., 1986), was a gift from Dr. Ryffel. The results from the cotransfections following treatment with DES, TX-OH, TPA, and EGF are shown in Figure 1. Figure 1 shows that diethylstilbestrol treatment in the presence of estrogen receptor results in high levels of CAT activity. This activity is comparable to constitutive CAT activity observed in the absence of receptor. In the presence of receptor and TX-OH, CAT activity is low. These results imply that the estrogen receptor acts as a repressor and that derepression is induced by the estrogenic ligand.

Figure 1 also shows that in the presence of receptor and DES although EGF has no effect on DES induced CAT activity, TPA attenuates the effect. This attenuation of CAT activity only occurs in the presence of receptor. Sucrose density gradient analysis of the receptor isolated from cells following transfection with CERO and treated with [^{3}H]estradiol ± TPA shows that receptors from control cells sediment at 3.5 S and that from cells treated with TPA

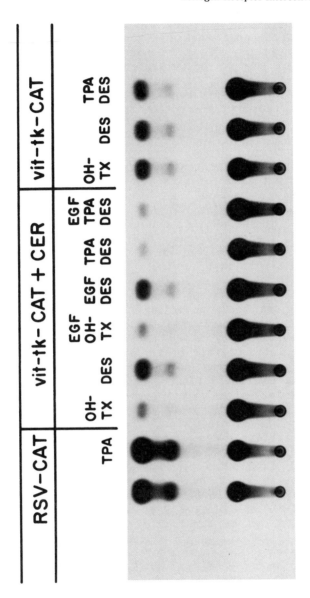

Figure 1. CAT activity in the cytosol of A431 cells transfected with RSV-CAT, pA2 (-821/-87) tk-CAT ± CER (chick estrogen receptor cDNA). The upper two spots are the acetylated forms of chloramphenicol.

sediment at 4.2 S (Nag et al., 1988). These experiments suggest that activation of the kinase C pathway results in maintenance of the estrogen receptor in the R_x conformation. The results also argue that the R_x conformation, which is also stabilized by antiestrogens (McNaught et al., 1986), might be capable of acting as a transcriptional repressor.

Experiments were also conducted using pA2 (−331/−87) tk−CAT instead of pA2 (−821/−87) tk−CAT. This construct also contains the ERE (Klein−Hitpass et al., 1986), however we were unable to see repression/derepression of CAT activity using this system. When the smaller construct lacking 490 bp of 5'−flanking region was used, DES clearly activated CAT activity above the constitutive level, but TPA did not inhibit this activity.

The observation that derepression versus activation can occur according to DNA sequences 5' of the ERE is particularly significant since it provides a dual mechanism by which estrogen responsive genes can be regulated differentially in the same cell. For example, activation of transcription involves recruitment of receptor−estrogenic ligand complexes to a particular DNA binding site; whereas, in derepression the receptor is already bound to a particular DNA binding site. Thus one would predict that the earliest response to estrogen treatment would be derepression, and later responses would probably be activated following increases in receptor concentration caused by ligand mediated increases in steroid receptor half−life and synthesis rates (Syms et al., 1985). While we have suggested that the R_x conformation can be involved in repression/depression according to DNA binding elements, we cannot exclude a role for R_{nb} since this receptor form was not assayed in these studies. If R_{nb} can also act as a repressor, the enzyme(s) involved in its conversion to R_y, would play an important regulatory role on estrogen mediation of gene transcription.

ISOLATION OF THE FACTOR WHICH CONVERTS R_{nb} TO R_y

In all our observations and experiments to date

R_y appeared to be crucial in activation of estrogenic events in vivo. The non-estradiol binding form of the receptor R_{nb} was also always present. Because of their association with estrogen mediated events we focused our attention on defining the factor(s) involved in the R_{nb} to R_y conversion process. Our initial experiments determined that the factor involved in this conversion, defined as F_y, could be separated from the receptors R_x, R_y and R_{nb} by ammonium sulfate fractionation (Dayani et al., 1988a). Interestingly, although we expected this factor to be a tyrosine kinase in agreement with earlier studies involving mammalian estrogen receptors from Auricchio's laboratory (Auricchio et al., 1981; Migliacchio et al., 1984; 1987), our partially purified preparation contained receptor phosphatase activity.

The above studies indicated that receptor dephosphorylation was associated with acquisition of steroid binding characteristics ie, conversion of R_{nb} to R_y (Dayani et al., 1988a). In more recent experiments F_y has been purified by DEAE-ion exchange chromatography, and size-exclusion HPLC. The conversion of R_{nb} to R_y using purified F_y is illustrated in Figure 2. Intriguingly F_y has been separated into kinase and phosphatase activities (McNaught et al., 1988; Dayani et al., 1988b). Both activities appear to be serine specific. The kinase has a molecular weight of approximately 40,000 and is ATP/Mg^{2+} dependent. Other cations and nucleotide substrates are inactive, and Ca^{2+} is not synergistic with Mg^{2+}. In the absence of this specific kinase or in the absence of the phosphatase the conversion of R_{nb} to R_y does not occur. Attempts to substitute exogenous kinases and phosphatases also did not allow the R_{nb} to R_y conversion (Dayani et al., 1988b). The F_y kinase and phosphatase appear to be unique in their substrate specificity. In particular the phosphatase appears to be most highly active on the estrogen receptor compared to other phosphorylated substrates.

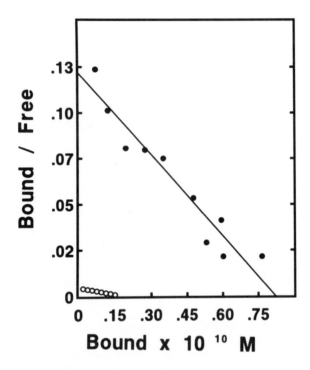

Figure 2. Scatchard Plot analysis of [³H] estradiol binding demonstrating the conversion of R_{nb} to R_y using purified F_y.   o-----o   30% ammonium sulfate purified receptor from chick oviduct cytosol isolated from chicks withdrawn from DES treatment for 48 hours; the cytosol was treated with 5mM Mg^{2+}/ATP at 30°/3h.   •-----•   same receptor preparation treated using the same protocol but in the presence of F_y.

It might seem curious at first that the conversion of R_{nb} to R_y requires a dephosphorylation event which is dependent upon ATP/Mg^{2+}. It is reasoned that F_y belongs to a class of protein phosphoryl phosphatases, thus there is a precedent for our observations (Ballou and Fischer, 1986). We suggest that in the absence of ATP/Mg^{2+} most of the phosphatase is in an inactive form. Following addition of ATP/Mg^{2+} the kinase F_y phosphorylates a regulatory or inhibitory subunit of the F_y phosphatase. This event causes a conformational change in the phosphatase resulting in its activation. This activated enzyme dephosphorylates R_{nb} resulting in the formation of R_y. In support of this notion, fluoride, vanadate, and pyrophosphate at concentrations known to inhibit serine phosphatases, inhibit the R_{nb} to R_y conversion. It appears therefore that in the chick oviduct a phosphoprotein phosphatase plays an important regulatory role in estrogen receptor mediated events.

CONCLUSIONS

The results of our reporter gene expression studies involving cotransfection of an estrogen receptor cDNA containing expression vector and vitellogenin-tk-CAT constructs, suggest that estrogen receptors can control gene expression by both repression/derepression and activation. It is speculated that flanking regions of the ERE element determine which of these events predominate and this provides a dual control mechanism for estrogen induced gene expression of different genes in the same cell. Since these conclusions are derived from studies in which the thymidine kinase promoter was used to allow constitutive expression of CAT activity, it might be argued that the system is artificial and not representative of gene activation in vivo. Alternatively, we would argue that it is perhaps typical of a relaxed chromatin structure typically associated with genes expressed as differentiation markers in estrogen target cells. The existence of tissue specific DNase sensitive sites associated with these genes is well documented. Certain of these sites remain sensitive even in the absence of active transcription; it is only particular DNase hypersensitive sites which change according to transcriptional status (Kaye et al., 1986). It is speculated therefore that in the absence of ligand, or in the presence of an antagonist, that genes which can be hormonally induced in target cells are repressed by the receptor because of its conformation upon binding to DNA. Binding of agonist alters receptor conformation and its phosphorylation state resulting in stabilization of the R_y form and relief of repression.

It is speculated that in one situation specific DNA sequences bind receptors with very high affinity in the absence of ligand. In this situation R_x, R_y and R_{nb} all bind and activation of transcription involves derepression. The alternative scenario is where the DNA sequences are such that they do not bind the receptor in the absence of agonist. In this situation stimulation of transcription involves direct activation rather than derepression. Our studies have indicated that both mechanisms are operable.

Future studies obviously require the characterization of DNA sequences surrounding the ERE which control receptor function. Ideally, for the appropriate DNA deletion studies, this requires the development of stable lines of A431 cells which express the chicken estrogen receptor. These studies are ongoing. Finally, a more complete characterization of the oviduct phosphoryl phosphatase and its regulation is required so that its role in estrogen mediated events can be defined more carefully.

ACKNOWLEDGEMENTS

NIH grant # HD17727 to RGS is gratefully acknowledged.

REFERENCES

Aurrichio F, Migliaccio G, Castoria S, Lastoria S, Schiavone E (1981). ATP dependent enzyme activating hormone binding of estradiol receptor. Biochem Biophys Res Commun 110:1171-1178.

Ballou LM, Fischer EH (1986). Phosphoprotein Phosphatases. In Boyer PD, Krebs EG (eds): The Enzymes: Academic Press, pp. 312–355.

Dayani N, McNaught RW, Smith RG (1988a). ATP mediated receptor interconversion as a model of estrogen action: Isolation of the factor which converts the non-estrogen binding form of the receptor to the lower affinity binding form. J Steroid Biochem 30:219–224.

Dayani, N, McNaught RW, Shenolikar S, Smith RG (1988b). Receptor interconversion model of hormone action: IV requirement of both kinase and phosphatase activities for conferring estrogen binding activity to the estrogen receptor. Submitted.

Gorman E (1985). "DNA cloning II." Glover DM (ed) IRL Press, Oxford 1985, pp 143–190.

Harris SE, Rosen JM, Means AR, O'Malley BW (1975). Use of a specific probe for ovalbumin mRNA to quantitate estrogen-induced gene transcripts. Biochemistry 14:2072–2081.

Hunter R, Ling N, Cooper J (1984). Protein kinase C phosphorylation of the EGF receptor at a threonine residue close to the cytoplasmic face of the plasma membrane. Nature 33:480–483.

Kaye JS, Pratt-Kaye S, Bellard M, Dretzen G, Bellard F, Chambon P (1986). Steroid hormone dependence of four DNase I-hypersensitive regions located within the 7000-bp 5'-flanking segment of the ovalbumin gene. EMBO Journal, 5:277–285.

Klein-Hitpass L, Schorpp M, Wagner U, Ryffel GU (1986). An estrogen-responsive element derived from the 5' flanking region of the Xenopus vitellogenin A2 gene functions in transfected human cells. Cell 46:1053–1061.

Lin CR, Chen WS, Lazar CS, Carpenter CD, Gill GN, Evans RM, Rosenfeld MG (1986). Protein kinase C phosphorylation at Thr654 of the unoccupied EGF receptor and EGF binding regulate functional receptor loss by independent mechanisms. Cell 44:839–848.

McNaught RW, Dayani N, Smith RG (1988). Receptor interconversion model of hormone action: III purification of a factor which confers estradiol binding properties to the estrogen receptor. Submitted.

McNaught RW, Smith RG (1986). Characterization of a second estrogen receptor in chick oviduct. Biochemistry 25:2073-2081.

McNaught RW, Raymoure WJ, Smith RG (1986). Receptor interconversion model of hormone action. I. ATP mediated conversion of estrogen receptors from a high to a lower affinity state and its relationship to antiestrogen action. J. Biol Chem 261:17011-17017.

Migliaccio A, Rotondi A, Aurrichio F (1984). Cal-modulin-stimulated phosphorylation of 17ß-estradiol receptor on tyrosine. Proc Natl Acad Sci 81:5921-5925.

Nag A, McNaught RW, Park I, Krust A, Smith RG (1988) Receptor interconversion model of hormone action V: estrogen receptor mediated repression of reporter gene activity in A431 cells. Submitted.

Raymoure RW, McNaught RW, Smith RG (1985). Reversible activation of non-steroid binding oestrogen receptor. Nature 314:745-747.

Raymoure RW, McNaught RW, Greene GL, Smith RG (1986). Receptors from non-steroid binding to the lower affinity binding state. J Biol Chem 261:17018-17025.

Sluyser M (1985). "Interaction of steroid hormone receptors with DNA." Sluyser M (ed.): Amsterdam, Netherlands.

Smith RG, Schwartz R (1979). Isolation and purifica-tion of the hen nuclear oestrogen receptor and its effect on transcription of chick chromatin. Biochem J 184:331-343.

Smith RG, Taylor RN (1981). Estrogen receptors as mediators of gene transcription. J Steroid Biochem 15:321-328.

Sutherland R, Mester J, Baulieu EE (1977). Tamoxifen is a potent pure antiestrogen in chick oviduct. Nature 267:434-435.

Syms A, Norris J, Panko W, Smith RG (1985). Mechanism of androgen receptor augmentation: Analysis of receptor synthesis and degradation by the density gradient shift technique. J Biol Chem 260:455-461.

Taylor RN, Smith RG (1982). Effects of highly purified estrogen receptors on gene transcription in isolated nuclei. Biochemistry 21:1781-1787.

Taylor RN, Smith RG (1985). Correlation in isolated nuclei of template-engaged RNA polymerase II, ovalbumin mRNA synthesis, and estrogen receptor concentrations. Biochemistry 24:1275–1280.

Theulaz I, Hipskind R, Heggeler-Bordier B, Green S, Kumar V, Chambon R, Wahli W (1988). Expression of human estrogen receptor mutants in Xenopus oocytes: correlation between transcriptional activity and ability to form protein – DNA complexes. EMBO J 7:1653–1660.

Molecular Endocrinology and Steroid
Hormone Action, pages 97–117
© 1990 Alan R. Liss, Inc.

EVIDENCE FOR A GLUCOCORTICOID RECEPTOR CYCLE AND NUCLEAR
DEPHOSPHORYLATION OF THE STEROID-BINDING PROTEIN.

Dirk B. Mendel, Eduardo Ortí, Lynda I. Smith,
Jack Bodwell, and Allan Munck.

Department of Physiology, Dartmouth Medical
School, Hanover, NH   03756

INTRODUCTION

An energy-dependent glucocorticoid receptor (GR)
cycle was proposed originally to account for the
observation that glucocorticoid binding capacity of cells
correlated with cellular ATP levels (Munck and
Brinck-Johnsen, 1968; Munck et al., 1972; Bell and Munck,
1973). Since ATP was not required for the GR to bind
hormone under cell-free conditions, the cyclic model
implied that phosphorylation was necessary for the
unliganded receptor to be functional. Similar receptor
cycles have since been proposed for other steroid hormone
receptors (Rossini, 1984).

Much indirect evidence, notably the demonstrations
that under cell-free conditions treatment of unoccupied
receptors with alkaline phosphatase caused loss of
steroid-binding capacity, and that spontaneous loss of
steroid-binding capacity could be reduced by phosphatase
inhibitors such as molybdate, supported the hypothesis
that phosphorylation of the receptor is necessary for
steroid binding (reviewed by Housley et al., 1984; Schmidt
and Litwack, 1982). Cyclic phosphorylation of the
receptor of course requires accompanying
dephosphorylation. Evidence with phosphatases and
phosphatase inhibitors has suggested that
dephosphorylation of the receptor occurs during the
activation process which converts the hormone receptor
complex to DNA-binding form (Barnett et al., 1980; Housley
et al., 1984; Matic and Trajkovic, 1986; Reker et al.,

1987).

     Direct evidence that the GR is phosphorylated in
intact cells was provided by Housley and Pratt (1983), who
showed that components of the unliganded receptor in
fibroblasts can be labeled with ^{32}P. We have extended
that result to WEHI-7 mouse thymoma cells, determining the
stoichiometry of phosphorylation (Mendel et al., 1987).
We are now looking systematically at the level of
phosphorylation of each functional form - nonactivated,
activated, nuclear bound, etc. - through which the
receptor passes in the cell.

     In this review we present direct evidence that there
is no dephosphorylation of the known components of the
nonactivated glucocorticoid receptor complex during
activation. We also show that the non-hormone-binding
form of the receptor present in ATP-depleted cells, the
'null' receptor, is partially dephosphorylated, and that
this dephosphorylation may be responsible for the tight,
but not covalent, association between the receptor and
nuclear matrix structures.

RESULTS

Evidence that activation does not involve
dephosphorylation of the steroid-binding or Hsp90
components of the nonactivated GR. In order to study the
relationship between receptor phosphorylation and receptor
function it is first necessary to be able to rapidly and
efficiently purify the various forms of the GR to near
homogeneity. In our studies we have used the BuGR1 and
BuGR2 monoclonal antibodies (Gametchu and Harrison, 1984)
to purify GR from WEHI-7 mouse thymoma cells.

     When purified nonactivated (non-DNA-binding)
cytosolic GRs are analyzed on polyacrylamide gels under
reducing and denaturing conditions (SDS-PAGE) we see two
prominent protein bands (Mendel et al., 1986a). One
protein migrates with a relative molecular mass of ~100
kDa, can be covalently labeled with the affinity steroid
[^{3}H]dexamethasone 21-mesylate (DM) (Simons and Thompson,
1981), and is specifically recognized by the BuGR
antibodies. We refer to this protein as the 100-kDa
steroid-binding protein. The other protein migrates with a

relative molecular mass of ~90 kDa, does not bind DM, and is not directly recognized by the BuGR antibodies (Mendel et al., 1986a). This protein has been referred to as the 90-kDa non-steroid-binding protein and it is copurified with the steroid-binding protein because these two proteins are associated through noncovalent interactions. Though it has been argued that the association between these two proteins is a cell-free artifact (reviewed in King, 1986), there is now evidence that this is not the case (Howard and Distelhorst, 1988a) and that these two proteins are associated in the intact cell (Rexin et al., 1988).

The 90-kDa non-steroid-binding component of nonactivated GR is common to other nonactivated steroid receptor complexes (Joab et al., 1984; Sullivan et al., 1985) and has been identified as the 90-kDa heat shock protein (Hsp90) (Catelli et al., 1985; Sanchez et al., 1985). Stoichiometric studies of the nonactivated GR have shown that it contains a single steroid-binding subunit (Okret et al., 1985; Gehring and Arndt, 1985) and two Hsp90 subunits (Mendel and Ortí, 1988). The 2 to 1 stoichiometric excess of Hsp90 in the nonactivated GR is consistent with the observation, based on the intensity of the stained bands following SDS-PAGE, that the Hsp90 is more abundant in purified GR complexes (Mendel et al., 1986a).

Analysis of purified nonactivated GR isolated from cells grown in medium containing $[^{32}P]$orthophosphate indicates that both the steroid-binding protein and Hsp90 are phosphoproteins (Housley et al., 1985; Mendel et al., 1986a). Both proteins are phosphorylated solely on serine residues (Dalman et al., 1988; Smith et al., 1988b) in agreement with the earlier results of Housley and Pratt (1983) and Kovacic-Milivojevic and Vedeckis (1986) using preparations in which the steroid-binding protein and Hsp90 were not resolved. It should be noted, however, that Auricchio et al. (1987) have reported that anti-phosphotyrosine antibodies can react with glucocorticoid receptors.

Since our primary interest was to study the phosphorylation state of various forms of the GR, it was first necessary to determine the phosphate content of the known components of the nonactivated complex so as to be sure that we could detect loss or gain of a single phosphate change. For the steroid-binding protein in unliganded GR isolated from WEHI-7 cells we have directly determined that there are, on average, 2.5 phosphates per steroid binding site (Mendel et al., 1987). With the demonstration that each steroid-binding protein contains a single steroid-binding site (Simons et al., 1987; Smith et al., 1988a; Carlstedt-Duke et al., 1988) we can now conclude that each steroid-binding protein contains either 2 or 3 phosphates.

A similar direct determination of the phosphate content of the Hsp90 was not possible since we had no way to directly measure the number of Hsp90 molecules being analyzed. However, from the incorporation of the ^{32}P label into the Hsp90 component of nonactivated complexes (Mendel et al., 1986a) and the stoichiometric composition of the antibody-purified nonactivated complex (Mendel and Ortí, 1988), we estimate that the two Hsp90 subunits associated with the nonactivated complex have a maximum of 5 or 6 total phosphates. This value is similar to the 5.8 phosphates reported for the Hsp90 dimer isolated from chick liver (Iannotti et al., 1988). Based on these determinations we are confident that we can detect the net loss of even a single phosphate from either the 100 kDa steroid-binding protein or the associated Hsp90.

In order to compare the phosphate content of the steroid-binding protein in the nonactivated with that in the activated hormone-receptor complex, WEHI-7 cells were grown in medium containing $^{32}P_i$ and [^{35}S]methionine. Following the labeling period (~18 hr), the cells were washed and resuspended in phosphate-free medium containing 200 nM triamcinolone acetonide (TA) at 37°C to form activated complexes in the intact cells. The cells were then lysed and the cytosol passed over a DNA-cellulose column to separate activated (bound to the column) from nonactivated (in the flow-through) complexes. The complexes were then isolated and analyzed on SDS-PAGE gels under conditions appropriate to resolve the two proteins (Mendel et al., 1986a). As a measure of phosphate content, the ^{32}P associated with the steroid-binding protein was normalized to the amount of ^{35}S incorporated, which was assumed to be proportional to the amount of the protein present, whether labeling with ^{35}S reached steady-state or not.

Table 1 shows that the relative phosphate contents of the steroid-binding protein from nonactivated and activated receptors is the same, indicating that this protein is not dephosphorylated during activation either in the intact cell (Mendel et al., 1987) or under cell-free conditions (Ortí et al., 1989). Since, as stated above, the steroid-binding protein in WEHI-7 cells contains few phosphates the net loss of a single phosphate would have been detected as a significant drop in the ratio of $^{32}P/^{35}S$ in the peak. Tienrungroj et al. (1987) using mouse fibroblasts subsequently arrived at a similar conclusion, although these authors did not determine the phosphate content of the steroid-binding protein in that system.

TABLE 1. Phosphate Content of the ~100-kDa Steroid-Binding Protein of Activated relative to Nonactivated GR.[*]

	Experiment	Relative Phosphate Content $\dfrac{\text{Activated Receptors}}{\text{Nonactivated Receptors}}$
Whole-cell Activation	1	1.0
	2	1.3
Cell-free Activation	3	0.9
	4	1.1
	5	0.8
Mean ± S.D.		1.02 ± 0.19

[*]Modified from Mendel et al., 1987; and Ortí et al., 1989.

Since the nonactivated GR complex is heteromeric, it is also possible that the Hsp90 component of the complex could undergo a phosphorylation-dephosphorylation cycle as it associates and dissociates from the steroid-binding protein. To test this possibility we have used the AC88 monoclonal antibody, which reacts with mammalian Hsp90 (Riehl et al., 1985), to purify free Hsp90 (i.e. the Hsp90 which is not associated with the receptor), and the BuGR antibodies to purify nonactivated GR from WEHI-7 cells grown in medium containing [^{35}S]methionine and ^{32}P$_i$. The BuGR-purified nonactivated GR complexes were activated by warming at 25°C for 30 min, and the Hsp90 which dissociated from the complexes was isolated. The free, receptor-associated, and dissociated forms of Hsp90 were then analyzed on SDS-PAGE gels and their relative phosphate content determined from the ratio of ^{32}P/^{35}S for the protein as described above for the steroid-binding protein.

Comparison of the relative phosphate contents of free Hsp90 with that of receptor-bound Hsp90 nonactivated complexes shows that they are the same (Table 2, middle column), indicating the Hsp90 is not phosphorylated as it becomes associated with the steroid-binding protein. Furthermore, as shown in the last column of Table 2, the Hsp90 is not dephosphorylated during cell-free activation (Ortí et al., 1989). It should be noted that activation had to be conducted using a cell-free system, since it would be impossible to distinguish between the Hsp90 which dissociated from the GR on activation and the vast excess of free Hsp in intact cells. However, unlabeled cytosol was added to the purified labeled receptors in these experiments to include any cytosolic components which might affect activation.

TABLE 2. Phosphate Content of Receptor-Bound Hsp90, and Hsp90 dissociated from the Receptor by activation, relative to Free Hsp90.*

| | Relative Phosphate Content* | |
| | Bound Hsp90 | Dissociated Hsp90 |
Experiment	Free Hsp90	Free Hsp90
1	0.92	0.96
2	1.02	1.20
3	0.98	1.26
4	0.91	1.09
Mean ± S.D.	0.96 ± 0.05	1.13 ± 0.13

*Modified from Ortí et al., 1989.

From these results, since we measure only net relative phosphate content of each protein, it is not possible to exclude the possibility of phosphate rearrangement during activation. However, in the experiments in which activation was carried out with purified receptors a phosphate rearrangement would have resulted in a net loss of phosphate, since no labeled ATP was present in the system.

Based on the results in Tables 1 and 2, we conclude that neither the steroid-binding protein nor the Hsp90 components of the nonactivated GR are dephosphorylated as part of the activation process. Given the ~330-kDa size of the WEHI-7 nonactivated complex under non-denaturing conditions (Sherman and Stevens, 1984; Holbrook et al., 1984), it is possible that this complex contains other components in addition to the single steroid-binding protein and the two Hsp90 subunits. RNA has been proposed to be associated with the activated complex (Kovacic-Milivojevic and Vedeckis, 1986; Ali and Vedeckis, 1987), and a 59-kDa protein has been reported to be a common component of a variety of nonactivated steroid-receptor complexes (Tai et al., 1986). Though we do not consistently find a 59-kDa protein we have not systematically looked for this protein and cannot rule out the possibility that it is lost during the purification process.

The apparent discrepancy between the lack of dephosphorylation of the receptor components during *in vivo* activation and the effects of phosphatases and phosphatase inhibitors could be explained if *in vitro* dephosphorylation destabilizes the oligomeric structure of the nonactivated receptor, resulting in dissociation of receptor subunits and enhanced activation of the complexes (Ortí et al., 1989).

We next tested to see whether the receptors were dephosphorylated in the nuclei of intact cells following activation and association with nuclear structures. Nuclear activated GR complexes can be distinguished by the conditions necessary to extract the complexes from isolated nuclei (Cidlowski and Munck, 1980). Most of the complexes, which we will refer to as the "loosely bound" nuclear GR, can be extracted with buffers containing 0.3 to 0.5 M NaCl or KCl, nucleases, or combinations of salt

and nucleases. After such extraction there still remains
a small fraction of the nuclear complexes, "tightly bound"
nuclear GR, which can only be extracted under denaturing
conditions. In WEHI-7 cells incubated with TA at 37 °C we
find that 10-20% of the nuclear activated complexes is
tightly bound, a figure which agrees with the value
reported by Gruol et al. (1983) for these same cells.

We have found that the loosely bound nuclear
activated complexes in WEHI-7 cells are phosphorylated to
the same extent as the cytosolic steroid-binding proteins
isolated from the same cells (Mendel et al., 1987). This
result, also reported by Tienrungroj et al. (1987) for
mouse L-cells, is not surprising in light of the fact that
the cytosolic activated and loosely bound nuclear
activated complexes appear to be in rapid equilibrium
(Miyabe and Harrison, 1983; Raaka and Samuels, 1983; Munck
and Holbrook, 1984).

Tightly bound nuclear activated complexes are thought
to include those complexes that generate biological
activity by association with glucocorticoid response
elements (Yamamoto and Alberts, 1976; Payvar et al.,
1981). The phosphorylation state of this form, is
therefore of particular interest to us. At this time,
however, we have not been able to satisfactorily determine
its relative phosphate, primarily because it represents
such a small fraction of the total cellular receptor.

Evidence that the non-hormone-binding 'null' receptor is
partially dephosphorylated. To determine whether a
dephosphorylated receptor could be induced in intact cells
we looked for the non-binding receptor postulated to exist
in ATP-depleted cells. Though there was indirect evidence
that this receptor is present in ATP-depleted cells (Munck
et al., 1972), it had not yet been identified. We
therefore looked for it in ATP-depleted WEHI-7 cells.

To our surprise, the non-hormone-binding form of the
receptor - which we called the glucocorticoid 'null'
receptor because of its inability to bind hormone - was in
the nuclear fraction of ATP-depleted WEHI-7 cells, even
though ATP-depletion was carried out in the absence of
hormone (Mendel et al., 1986b). We have subsequently
characterized this null receptor. A summary of the
findings of that study are presented in Table 3.

TABLE 3. Characteristics of Glucocorticoid Null Receptors.

---

1. Null receptors are formed by various treatments which lower cellular ATP levels.

2. Null receptors do not appear to be oxidized forms of the receptor. They are not disulfide-linked to nuclear components. Under ATP-depletion total sulfhydryls remain unchanged but binding capacity drops to 30% of controls.

3. Null receptors are rapidly reconverted to cytosolic receptors capable of binding hormone upon the restoration of cellular ATP. This process does not appear to require protein synthesis and therefore is unlikely to be the result of receptor synthesis.

4. Null receptors are tightly associated with nuclear structures. They are resistant to extraction with salt and nuclease concentrations which extract most nuclear activated complexes from cells which have not been depleted of ATP.

Of particular interest among these results is the observation that null receptors are extremely tightly associated with nuclear structures. In fact, as is shown in Fig. 1, null receptors cannot be extracted with buffers containing elevated salt concentrations even in the presence of nucleases and reducing agents. In this respect null receptors behave like the tightly bound nuclear activated complexes which are formed in cells under normal physiologic conditions. This suggests that null receptors may be a model for tightly bound nuclear activated complexes.

Recently Kaufmann et al. (1986) reported that nuclear activated complexes can become linked to nuclear matrix proteins via disulfide bond formation if the nuclei are prepared in the absence of sulfhydryl protecting agents. Under conditions which favor disulfide bond formation, most of the activated nuclear GR become resistant to extraction even by buffers containing 1.6 M NaCl and nucleases. However, since the GR are disulfide-linked to

nuclear matrix proteins under these conditions, they can
be readily extracted if reducing agents are included in
the extraction buffers.

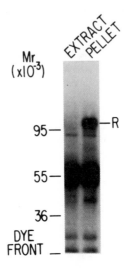

Figure 1. Extraction of nuclei containing null receptors
with nucleases, salt and reducing agent.
        WEHI-7 cells were incubated for 90 min at 37 °C in
the presence of 1 mM 2,4-dinitrophenol to generate null
receptors. Following cell lysis ~10[8] nuclei were incubated
with 1 mg/ml DNase I and 1 mg/ml RNase A, at 4 °C for 60
min (under these conditions extraction of nuclear
activated [3H]TA-receptor complexes is maximal). Nuclei
were further extracted for 30 min in the presence of 1.6 M
NaCl and 10 mM DTT. After centrifugation (11,000 x g, 5
min) the pellet was dissolved by heating to 100 °C in the
presence of 4 % SDS. GR were purified from the extract and
the solubilized pellet by immunoabsorbtion using the BuGR2
antibody, and detected by Western blotting following
SDS-PAGE using BuGR2 as the first antibody and
[125I]anti-mouse IgG as the second antibody. MW markers
and the receptor (R) band are indicated. The intense
labeling at ~55 kDa corresponds to BuGR2 antibody protein.

        An obvious question which comes out of our
observation that null receptors are resistant to

extraction with salt and nucleases is whether this receptor form is linked by disulfides to nuclear matrix structures. Two lines of evidence argue against such disulfide linkages. First, null receptors cannot be extracted by high salt and nucleases either in the absence or presence of reducing agents. This contrasts with the result found by Kaufmann et al. (1986). In the experiment shown in Fig. 1 we used 10 mM DTT as the reducing reagent, but even when we used 100 mM mercaptoethanol to insure that sufficient reducing agent was present we obtained the same result.

Our second line of evidence comes from studies in which we have analyzed null receptors, extracted from the nuclei of ATP-depleted cells under denaturing conditions in the absence of reducing agents, on non-reducing SDS-PAGE gels. We find that even under these conditions the null receptor migrates primarily as a ~100-kDa protein (D.B.M., E.O., A.M., manuscript in preparation), as does the cytosolic receptor isolated from cells under normal metabolic conditions. We do not see the large aggregated complexes reported by Kaufmann et al. (1986) which is apparently due to disulfide linkages between the GR and nuclear matrix proteins.

At present we do not know how null receptors are bound to nuclear structures or whether they are associated with glucocorticoid response elements. Our assumption that they are associated with nuclear matrix structures is based on the observation that null receptors are associated with a nuclear pellet following digestion of nuclei under conditions generally used to isolate the nuclear matrix (Kaufmann et al., 1986).

Having partially characterized the null receptors, we then compared the phosphate content of null receptors with that of cytosolic receptors from cells which had not been depleted of ATP. Essentially as described above for the studies in which we determined the phosphate content of the steroid-binding protein and Hsp90, WEHI-7 cells were grown for 18-24 hr in medium containing $^{32}P_i$ and [35S]methionine to dual label the cellular proteins. The cells were harvested, washed free of glucose and unincorporated label, and resuspended in glucose-free phosphate-free medium. The cell suspensions were then split into 2 or 3 equal aliquots. Either DNP or antimycin

A was added to cause a reduction in cellular ATP, and the rest of the cell suspension was incubated under the same condition but without addition of either agent. We previously showed that without inhibitors, ATP levels in WEHI-7 cells do not drop over the 90-min incubation period, and the binding capacity is largely retained (Mendel et al., 1986b). At the end of the 90-min incubation the cells were broken and the steroid-binding protein antibody-purified from the cytosolic (in the case of the untreated cells) or nuclear (in the case of the ATP-depleted cells) fraction of the cells. Following SDS-PAGE, the phosphate content (^{32}P counts) relative to the amount of steroid-binding protein (^{35}S counts) was determined for the ~100-kDa steroid-binding protein isolated from each group of cells.

From the results of 3 separate experiments, shown in Table 4, it is clear that the phosphate content of the steroid-binding protein of null receptors is less than that of unoccupied complexes isolated from cells which had not been depleted of ATP. It is also clear that this decrease occurs whether ATP is depleted by DNP or by antimycin A, an agent which depletes ATP through a mechanism different from that of DNP.

The observation that the phosphate content of the steroid-binding protein isolated from ATP-depleted cells is 65 % of that found in control cells is consistent with our interpretation that null receptors have lost one of the three total phosphates originally associated with the receptor. Whether the loss of this phosphate is responsible for the loss of steroid binding capacity remains to be determined. In any case this is the first example of a partly dephosphorylated form of the glucocorticoid receptor isolated from intact cells, and supports the hypothesis that dephosphorylation may be part of a receptor cycle.

It is tempting to speculate that the partial dephosphorylation of the receptor is responsible for the receptor becoming tightly associated with the nuclear matrix. This would explain data from our own (Mendel et al., 1986b) and from other laboratories (Ishii et al., 1972) suggesting that release of the receptor from the nucleus is an energy-dependent process. Whether this provides a means for regulating transcriptional activity of the receptor also remains to be determined. It has been reported that the activity of transcription factors may depend on the extent of their phosphorylation (reviewed in Ptashne, 1988).

TABLE 4. Phosphate content of null receptors relative to receptors isolated from the cytosol of cells incubated under normal conditions

Experiment	Agent used for ATP-Depletion	Relative Phosphate Content $\dfrac{\text{Null Receptors}}{\text{Cytosolic Receptors}}$
1	DNP	0.67
2	DNP	0.57
	Antimycin	0.70
3	DNP	0.60
	Antimycin	0.73
Mean ± S.D.		0.65 ± 0.07

We have not yet identified which phosphate is lost during null receptor formation. Dalman et al. (1988) have reported that in mouse fibroblast L-cells there is a phosphorylation site within the chymotryptic receptor fragment which contains the DNA-binding and steroid-binding domains of the receptor, and conclude that the phosphate is in the DNA-binding domain. From our own studies with WEHI-7 cells (Smith et al., 1988b) we also find that there is a single phosphorylation site within the chymotryptic fragment of the steroid-binding protein. However, this phosphate appears to be associated with the ~29-kDa tryptic receptor fragment which contains the steroid-binding domain, not with the ~16-kDa tryptic fragment of the receptor which contains both the BuGR-reactive antigen and the DNA-binding domain. Presumably the remaining phosphates are in the amino half of the molecule which has been called the immunogenic domain. Based on these results, we think that the

phosphate in the steroid binding domain is the most likely
to have been lost in the null receptor. Support for this
view comes from the fact that the nt[1] mutant receptor,
which lacks the immunogenic domain (Northrop et al.,
1985), and a 50-kDa proteolytic fragment of the rat thymus
receptor, which contains the steroid- and DNA-binding
domains and therefore presumably also lacks an immunogenic
domain (Mendel et al., 1985), can both bind hormone.
Since the immunogenic domain and its phosphates are
unnecessary for steroid binding, whatever is in the
steroid- and DNA-binding domains is sufficient to bind
hormone.

CONCLUSIONS

From the results presented here we propose the
following updated model of the putative glucocorticoid
receptor cycle, with special emphasis on receptor subunit
composition and phosphorylation (Fig. 2). The left hand
portion of this model depicts the binding of the hormone
(H) to the cytosolic unliganded receptor (R) to form the
nonactivated complex (HR), activation of HR to the
activated complex (HR') with dissociation of the Hsp90
subunits from the steroid-binding protein, and the
association of HR' complex with nuclear structures to form
loosely bound nuclear activated complexes (HR'n). These
reactions have been demonstrated by a variety of
laboratories to occur under cell-free conditions as well
as in intact cells. The phosphorylation state of these
complexes have been directly determined (Mendel et al.,
1987) and there is no change in the net phosphate content
of the steroid-binding protein or the associated Hsp90's
during activation or loose nuclear binding (Mendel et al.,
1987; Tienrungroj et al., 1987; Ortí et al., 1989).

The right hand portion of this model, which shows the
conversion of HR'n to tightly bound nuclear complexes
(HR"n) and the regeneration of functional cytosolic
unliganded receptors (R) which can bind a different
glucocorticoid molecule, is at present hypothetical and
therefore indicated with broken lines. The fact that the
null receptor characteristics are very similar to those of
the 10-20 % nuclear activated complexes which are also
tightly bound (HR"n) suggests to us that this form of the
receptor may also be partially dephosphorylated. In the

model we show loss of a phosphate when HR'n is converted to HR"n, and dissociation of hormone from this latter form. The unoccupied tighly bound receptor (R"n) is then assumed to be released from the nucleus through an ATP-dependent process, presumably involving rephosphorylation of the steroid binding protein.

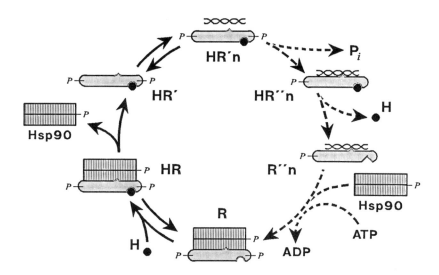

Figure 2. Cyclic model for glucocorticoid receptor function in intact cells.

R, unliganded nonactivated receptor; HR, nonactivated hormone receptor complex; HR', activated hormone receptor complex; HR'n, loosely bound nuclear activated hormone-receptor complex; HR"n, tightly bound nuclear activated hormone receptor complex; R"n, tightly bound nuclear unliganded receptor. Phosphates (P) are indicated generically, without relation to actual numbers or locations.

We have included as part of this step the association between the steroid-binding protein and the two molecules of Hsp90 to form the oligomeric unoccupied receptor R. What is required for these two proteins to associate is uncertain, since to date reconstitution of R from steroid-binding protein and Hsp90 has not been achieved. The role the Hsp90 plays is still unclear, but it has been suggested that it is required to maintain steroid-binding capability of the steroid-binding protein (Bresnick et al., 1988) and to prevent the receptor from binding to the genome (Baulieu, 1987; Pratt, 1987; Howard and Distelhorst, 1988).

The formation of null receptors in ATP-depleted cells can be explained on the basis of the proposed model by assuming that unliganded receptors are able to undergo similar cyclic processes but at a slower rate than agonist-bound receptors. Since in the absence of ATP the rephosphorylation is blocked, the dephosphorylated form of the receptor (R"n) accumulates in the nucleus. Such a slow hormone-independent cycle would imply that unliganded receptors might produce a low level of biological activity.

Though the formation of null receptors occurs under conditions which are not physiological, this form of the receptor may be useful for studying mechanisms involved in receptor phosphorylation in intact cells, and has potential as a model for tightly bound forms of the receptor which are present in low amounts in intact cells.

ACKNOWLEDGEMENTS

The studies described here were supported by Research Grant DK 03535 from the U.S. Public Health Service and by the Core Grant of the Norris Cotton Cancer Center (CA 23108). D.B.M. held a Cancer Research Institute/Miriam and Benedict Wolf Fellowship; E.O. was supported by fellowships from the Consejo Nacional de Investigaciones Científicas y Técnicas de la República Argentina and from the U.S. Public Health Service (1F05 TW03923-01); and L.I.S. received a predoctoral fellowship from the Endocrine Training Grant DK 07508.

REFERENCES

Ali M, Vedeckis WV (1987). Interaction of RNA with the transformed glucocorticoid receptor. J Biol Chem 262:6771-6777.

Auricchio F, Migliaccio A, Castoria G, Rotondi A, Di Domenico M, Pagano M, Nola E (1987). Phosphorylation on tyrosine of oestradiol-17β receptor in uterus and interaction of oestradiol-17β and glucocorticoid receptors with antiphosphotyrosine antibodies. J Steroid Biochem 27:245-253.

Barnett CA, Schmidt TJ, Litwack G (1980). Effects of calf intestinal alkaline phosphatase, phosphatase inhibitors and phosphorylated compounds on the rate of activation of glucocorticoid-receptor complexes. Biochemistry 19:5446-5455.

Baulieu EE (1987). Antihormone-steroid hormonal activity, heat-shock protein hsp 90 and receptors. Hormone Res 28:181-195.

Bell PA, Munck A (1973). Steroid-binding properties and stabilization of cytoplasmic glucocorticoid receptors from rat thymus cells. Biochem J 136:97-107.

Bresnick EH, Sanchez ER, Pratt WB, (1988). Relationship between glucocorticoid receptor steroid-binding capacity and association of the M_r 90,000 heat shock protein with the unliganded receptor. J Steroid Biochem 30:1-6.

Catelli MG, Binart N, Jung-Testas I, Renoir JM, Baulieu EE, Feramisco JR, Welch WJ (1985). The common 90-kd protein component of non-transformed "8S" steroid receptors is a heat shock protein. EMBO J 4:3131-3135.

Carlstedt-Duke J, Strömstedt P-E, Persson B, Cederlund E, Gustafsson J-A, Jörnvall H (1988). Identification of hormone-interacting amino acid residues within the steroid-binding domain of the glucocorticoid receptor in relation to other steroid hormone receptors. J Biol Chem 263:6842-6846.

Cidlowski JA, Munck A (1980). Multiple forms of nuclear binding of glucocorticoid-receptor complexes in rat thymocytes. J Steroid Biochem 13:105-112.

Dalman FC, Sanchez ER, Lin AL-Y, Perini F, Pratt WB (1988). Localization of phosphorylation sites with respect to the functional domains of the mouse L cell glucocorticoid receptor. J Biol Chem 263:12259-12267.

Gametchu B, Harrison RW (1984). Characterization of a monoclonal antibody to the rat liver glucocorticoid receptor. Endocrinology 114:274-279.

Gehring U, Arndt H (1985). Heteromeric nature of glucocorticoid receptors. FEBS Letters 179:138-142.

Gruol DJ, Kempner ES, Bourgeois S (1984).
Characterization of the glucocorticoid receptor -
comparison of wild type and variant receptors. J Biol
Chem 259:4833-4839.

Holbrook NJ, Bodwell JE, Munck A (1984). Nonactivated and
activated glucocorticoid-receptor complexes in WEHI-7
and rat thymus cells. J Steroid Biochem 20:19-22.

Housley PR, Pratt WB (1983). Direct demonstration of
glucocorticoid-receptor phosphorylation by intact L
cells. J Biol Chem 258:4630-4635.

Housley PR, Grippo JF, Dahmer MK, Pratt WB (1984).
Inactivation, activation, and stabilization of
glucocorticoid receptors. Biochemcal Actions of
Hormones 11:347-376.

Housley PR, Sanchez ER, Westphal HM, Beato M, Pratt WB
(1985). The molybdate-stabilized L-cell glucocorticoid
receptor isolated by affinity chromatography or with a
monoclonal antibody is associated with a 90-92 kDa
nonsteroid-binding phosphoprotein. J Biol Chem
260:13810-13817.

Howard KJ, Distelhorst CW (1988a). Evidence for
intracellular association of glucocorticoid receptor
with the 90-kDa heat shock protein. J Biol Chem
263:3474-3481.

Howard KJ, Distelhorst CW (1988b). Effect of the 90 kDa
heat shock protein, hsp90, on glucocorticoid receptor
binding to DNA-cellulose. Biochem Biophys Res Comm
151:1226-1232.

Ishii DN, Pratt WB, Aronow L (1972). Steady-state level
of the specific glucocorticoid binding component in
mouse fibroblasts. Biochemistry 11:3896-3904.

Iannotti AM, Rabideau DA, Dougherty JJ (1988).
Characterization of purified avian 90,000-Da heat shock
protein. Arc Biochem Biophys 264:54-60.

Joab I, Radanyi C, Renoir M, Buchou T, Catelli MG, Binart
N, Mester J, Baulieu EE (1984). Common non-hormone
binding component in non-transformed chick oviduct
receptors of four steroid hormones. Nature
308:850-953.

Kaufmann SH, Okret S, Wikström A-C, Gustafsson J-A, Shaper
JH (1986). Binding of the Glucocorticoid receptor to
the rat liver nuclear matrix. J Biol Chem
261:11962-11967.

King RJB (1986). Receptor structure: A personal
assessment of the current status. J Steroid Biochem
25:451-454.

Kovacic-Milivojevic B, Vedeckis WV (1986). Absence of detectable ribonucleic acid in the purified, untransformed mouse glucocorticoid receptor. Biochemistry 25:8266-8273.

Matic G, Trajkovic D (1986). The effect of alkaline phosphatase on the activation of glucocorticoid-receptor complexes in rat liver cytosol. J Biochem 18:841-845.

Mendel DB, Holbrook NJ, Bodwell JE (1985). Degradation without apparent change in size of molybdate-stabilized nonactivated glucocorticoid-receptor complexes in rat thymus cytosol. J Biol Chem 260:8736-8740.

Mendel DB, Bodwell JE, Gametchu B, Harrison RW, Munck A (1986a). Molybdate-stabilized nonactivated glucocorticoid-receptor complexes contain a 90-kDa non-steroid-binding phosphoprotein that is lost on activation. J Biol Chem 261:3758-3763.

Mendel DB, Bodwell JE, Munck A (1986b). Glucocorticoid receptors lacking hormone-binding activity are bound in nuclei of ATP-depleted cells. Nature 324:478-480.

Mendel DB, Bodwell JE, Munck A (1987). Activation of cytosolic glucocorticoid-receptor complexes in intact WEHI-7 cells does not dephosphorylate the steroid-binding protein. J Biol Chem 262:5644-5648.

Mendel DB, Ortí E (1988). Isoform composition and stoichiometry of the ~90-kDa heat shock protein associated with glucocorticoid receptors. J Biol Chem 263:6695-6702.

Miyabe S, Harrison RW (1983). *In vivo* activation and nuclear binding of the AtT-20 mouse pituitary tumor cell glucocorticoid receptor. Endocrinology 112:2174-2180.

Munck A, Brinck-Johnsen T (1968). Specific and nonspecific physicochemical interactions of glucocorticoids and related steroids with rat thymus cells in vitro. J Biol Chem 243:5556-5565.

Munck A, Wira C, Young DA, Mosher KM, Hallahan C, Bell PA (1972). Glucocorticoid-receptor complexes and the earliest steps in the action of glucocorticoids on thymus cells. J Steroid Biochem 3:567-578.

Munck A, Holbrook NJ (1984). Glucocorticoid-receptor complexes in rat thymus cells. Rapid kinetic behavior and a cyclic model. J Biol Chem 259:820-831.

Northrop JP, Gametchu B, Harrison RW, Ringold GM (1985).
Characterization of wild-type and mutant glucocorticoid
receptors from rat hepatoma and mouse lymphoma cells.
J Biol Chem 260:6398-6403.

Okret S, Wikstrom AC, Gustafsson JA (1985).
Molybdate-stabilized glucocorticoid receptor: Evidence
for a receptor heteromer. Biochemistry 24:6581-6586.

Ortí E, Mendel DB, Munck A (1989). Phosphorylation of
glucocorticoid receptor-associated and free forms of
the ~90-kDa heat shock protein before and after
receptor activation. J Biol Chem 264:231-237.

Payvar F, Wrange Ö, Carlstedt-Duke J, Okret S, Gustafsson
JA, Yamamoto KR (1981). Purified glucocorticoid
receptors bind selectively in vitro to a cloned DNA
fragment whose transcription is regulated by
glucocorticoids in vivo. PNAS 78:6628-6632.

Pratt WB (1987). Transformation of glucocorticoid and
progesterone receptors to the DNA-binding state. J
Cell Biochem 35:51-68.

Ptashne M (1988). How eukaryotic transcriptional
activators work. Nature 335:683-689.

Raaka BM, Samuels HH (1983). The glucocorticoid receptor
in GH_1 cells. J Biol Chem 258:417-425.

Reker CE, LaPointe MC, Kovacic-Milivojevic B, Chiou WJH,
Vedeckis WV (1987). A possible role for
dephosphorylation in glucocorticoid receptor
transformation. J Steroid Biochem 26:653-665.

Rexin M, Busch W, Gehring U (1987). Chemical
cross-linking of heteromeric glucocorticoid receptors.
Biochemistry 27:5593-5601.

Riehl RM, Sullivan WP, Vroman BT, Bauer VJ, Pearson GR,
Toft DO (1985). Immunological evidence that the
nonhormone binding component of avian steroid receptors
exists in a wide range of tissues and species.
Biochemistry 24:6586-6591.

Rossini GP (1984). Steroid receptor recycling and its
possible role in the modulation of steroid hormone
action. J Theor Biol 108:39-53.

Sanchez ER, Toft DO, Schlesinger MJ, Pratt WB (1985).
Evidence that the 90 kDa phosphoprotein associated with
the untransformed L-cell glucocorticoid receptor is a
murine heat shock protein. J Biol Chem
260:12398-12401.

Schmidt TJ, Litwack G. (1982). Activation of the
glucocorticoid-receptor complex. Physiol Rev
62:1131-1192.

Sherman MR, Stevens J (1984). Structure of mammalian steroid receptors: Evolving concepts and methodological developments. Ann Rev Physiol 46:83-105.

Simons SS, Thompson EB (1981). Dexamethasone 21-mesylate: An affinity label of glucocorticoid receptors from rat hepatoma tissue culture cells. Proc Natl Acad Sci USA 78:3541-3545.

Simons SS, Pumphrey JG, Rudikoff S, Eisen HJ (1987). Identification of cysteine 656 as the amino acid of hepatoma tissue culture cell glucocorticoid receptors that is covalently labeled by dexamethasone 21-mesylate. J Biol Chem 262:9676-9680.

Smith LI, Bodwell JE, Mendel DB, Ciardelli T, North WG, Munck A (1988a). Identification of cysteine-644 as the covalent site of attachement of dexamethasone 21-mesylate to murine glucocorticoid receptors in WEHI-7 cells. Biochemistry 27:3747-3753.

Smith LI, Mendel DB, Bodwell JE, Munck A (1988b). Phosphorylated sites within the functional domains of the 100-kDa steroid binding subunit of glucocorticoid receptors (submitted for publication).

Sullivan WP, Vroman BT, Bauer VJ, Puri RK, Riehl RM, Pearson GR, Toft DO (1985). Isolation of steroid receptor binding protein from chicken oviduct and production of monoclonal antibodies. Biochemistry 24:4214-4222.

Tai PK, Maeda Y, Nakao K, Wakim NG, Duhring JL, Faber LE (1986). A 59-kilodalton protein associated with progestin, estrogen, androgen, and glucocorticoid receptors. Biochemistry 25:5269-5275.

Tienrungroj W, Sanchez ER, Housley PR, Harrison RW, Pratt WB (1987). Glucocorticoid receptor phosphorylation, transformation, and DNA binding. J Biol Chem 262:17342-17349.

Yamamoto KR, Alberts BM (1976). Steroid receptors: Elements for modulation of eukaryotic transcription. Ann Rev Biochem 45:721-746.

Molecular Endocrinology and Steroid
Hormone Action, pages 119–132
© 1990 Alan R. Liss, Inc.

GLUCOCORTICOID RECEPTOR STRUCTURE AND THE INITIAL EVENTS IN
SIGNAL TRANSDUCTION

William B. Pratt

Department of Pharmacology, University of
Michigan Medical School, Ann Arbor Michigan 48109

INTRODUCTION

The binding of steroid to a receptor in an intact cell
initiates a change in the receptor such that it is then
able to participate in the next step in the chain of events
that result in hormone-regulated gene transcription. This
initial steroid-mediated change in the state of the
receptor has been called receptor "transformation" or
"activation". Estrogen or progesterone receptors are
located predominantly in the nucleus prior to receptor
transformation. Upon binding steroid, the receptors are
converted in a temperature-dependent manner from a form
that readily dissociates from nuclear structures (i.e., it
is recovered in the cytosolic fraction of the cell) to a
form that is tightly associated with an as yet undefined
nuclear structure. In contrast to estrogen and
progesterone receptors, glucocorticoid receptors are
located predominantly in the cytoplasm in the absence of
hormone. Upon binding glucocorticoid, the receptor rapidly
moves to the nucleus (Wikstrom et al., 1987; Picard and
Yamamoto, 1987) where it is bound in a high affinity
manner.

To understand the way in which the free energy
involved in steroid binding is transduced into a
transformation of the receptor from an inactive to an
active state, we must eventually define the structure of
the untransformed and transformed states of the receptor in
the cell. In this chapter, I will review some recent

experiments with glucocorticoid receptors that have advanced our understanding of the hormone-mediated transformation event. Specifically, I will develop the proposal that, prior to exposure to hormone, the receptor is associated with a structure responsible for its transport to the nucleus and perhaps through the nuclear pores. This structure contains the abundant 90-kDa heat shock protein, hsp90, and the receptor-hsp90 complex may be transported to the nucleus along a scaffold composed of microtubules. Binding of glucocorticoids permits a temperature-dependent dissociation of hsp90 from the receptor with the resulting exposure of both the DNA-binding domain and one or more additional domains involved in transfer of the receptor from the cytoplasm to nucleus and in the ultimate localization of the receptor in its high affinity nuclear site of action.

RECEPTOR TRANSFORMATION IN CYTOSOL PREPARATIONS

About 10 years ago, it was shown that molybdate stabilizes cytosolic steroid receptors in a steroid binding form and inhibits their transformation from a non-DNA-binding to a DNA-binding form. This transformation of receptors to the DNA-binding state in cytosol is both temperature-dependent and hormone-dependent and has been studied as a cell-free model of receptor transformation in the intact cell (Sanchez et al., 1987). In order to study the transformation process, several laboratories exploited the stabilizing effect of molybdate to facilitate purification of steroid receptors in their untransformed, non-DNA-binding state. It was shown that untransformed glucocorticoid, progesterone, estrogen and androgen receptors are associated with a non-steroid-binding phosphoprotein which has been identified as hsp90. Transformed receptors in cytosol are not associated with the heat shock protein (for reviews see Pratt, 1987; Pratt et al., 1989).

These observations led to the concept that transformation of cytosolic glucocorticoid receptors resulted from the dissociation of a heteromeric receptor-hsp90 complex (Sanchez et al., 1985). It was subsequently shown that dissociation of hsp90 from cytosolic glucocorticoid receptor is both a hormone-dependent and a temperature-dependent event in the same manner as generation of the DNA-binding state (Sanchez et al., 1987). The receptor-hsp90 complex, with a probable stoichiometry of 2 molecules of hsp90 to one molecule of steroid binding protein (Mendel and Orti, 1988), is the 9S form of the receptor and the dissociated DNA-binding form is the 4S receptor.

From cross-linking experiments, it is now well established that the glucocorticoid receptor is bound to hsp90 in intact cells under hormone-free conditions (Rexin et al., 1988). Recent genetic evidence suggests that the 9S cytosolic GR is derived from the normal untransformed state of the receptor in the cell (Pratt et al., 1988). It was reasoned that if the 9S form of the receptor is related to repression of receptor function in the intact cell, then there should be correlations between the biological properties (i.e., constitutive vs. inducible) of receptors produced from modified receptor cDNAs and their recovery in cytosol as 9S complexes. Accordingly, COS cells were transfected with wild-type and mutant human glucocorticoid receptor cDNAs and the sedimentation behavior of the receptors was determined after preparing cytosol from the transfected cells. We found that the steroid-inducible forms of the receptor were all recovered in the 9S form, whereas the mutants with constitutive activity were recovered only in the 4S form. This observation strongly supports the proposal that the 9S heteromeric complex is derived from the inactive form of the receptor in the intact cell. It follows also that conversion of receptors from 9S to 4S and the accompanying conversion from a non-DNA-binding to a DNA-binding form in cytosol represents a biologically meaningful cell-free model system for studying the critical first step in signal transduction by glucocorticoids.

THE STEROID BINDING DOMAIN DETERMINES BOTH REPRESSION OF RECEPTOR FUNCTION AND FORMATION OF THE 9S COMPLEX

From transfection studies, it has been determined that features in the steroid binding domain of the glucocorticoid receptor normally repress receptor function in intact cells (e.g., Hollenberg et al., 1987). In cytosols, the 9S receptor is repressed in the sense that the DNA binding function is blocked. It is clear that the structural features required for formation of the 9S complex and association with hsp90 are located within the steroid binding domain (Pratt et al., 1988; Denis et al., 1988). The steroid binding domains of the steroid receptors possess amino acid sequences that differ widely between different classes of steroids. This diversity is undoubtedly required to determine specificity of hormone binding. It is, however, reasonable to predict that the mechanism by which steroid binding is transduced into a derepression of receptor function would be conserved among receptors for the different steroid classes. As the steroid-binding domain represses the function of the DNA-binding domain, it is also reasonable to propose that the region responsible for steroid-mediated derepression of the DNA-binding domain would be determined by a conserved sequence lying near the steroid binding site. Figure 1 shows a 20 amino acid region of high homology existing within the steroid binding domains of all of the steroid

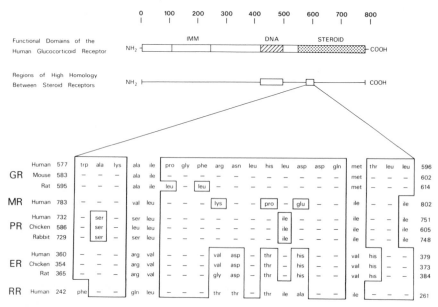

Figure 1. A potential "transducing structure" lying within the hormone binding domain of steroid receptors.

receptors. This domain may form a structure or part of a structure that is responsible for transducing the free energy of steroid binding into derepression of receptor function. In the simplest model, which is shown in Figure

2, the transducing domain would form the binding site for high affinity interaction with hsp90, which overlays or in some other way represses the DNA-binding function.

The model of Figure 2 takes into account several known functional domains and topological features of the mouse glucocorticoid receptor. The locations of the epitope for the BuGR monoclonal antibody (Rusconi and Yamamoto, 1987), the endogenous phosphorylation sites (Dalman et al., 1988), and the DNA-binding domain (Danielsen et al., 1986) in the primary sequence of the mouse receptor are established. The steroid binding pocket is determined by the carboxy terminal 25-kDa segment of the receptor. Cys644 is covalently labeled by the glucocorticoid antagonist dexamethasone 21-mesylate (Simons et al., 1987; Smith et al., 1988) and both met610 and cys742 are labeled by

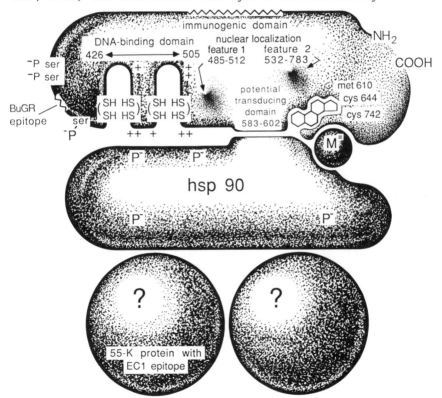

Figure 2. Model of the untransformed, non-DNA-binding form of the murine cytosolic glucocorticoid receptor.

triamcinolone acetonide after u.v. irradiation (Carlstedt-Duke et al., 1988). It is important to note that met610 lies right next to the potential transducing domain outlined in Figure 1. In the simplest version of the model, the binding of steroid permits a temperature-dependent event that leads to a reduction in the affinity of this site for hsp90 and consequent derepression of the DNA-binding domain. The globe with the M represents either molybdate or an endogenous metal anion, both of which stabilize the hsp90-receptor complex and thus prevent transformation to the DNA-binding state (Meshinchi et al., submitted manuscript).

SPECULATIONS REGARDING THE INITIAL EVENTS IN GLUCOCORTICOID-MEDIATED SIGNAL TRANSDUCTION

Our working model of the initial event in steroid-mediated signal transduction is that the free energy involved in the binding of steroid to its high affinity binding site is transduced through a change in the conformation of the conserved "transducing domain" to yield a substantial increase in the dissociation rate constant for the hsp90-"transducing domain" complex. The interaction between the hsp90 binding site and the steroid binding site may explain another phenomenon as well. It has been known for a number of years that glucocorticoids do not bind to the dissociated 4S form of the receptor, although they readily bind to the isolated 9S complex. Recently, Bresnick et al. (submitted manuscript) have made direct quantitative measurements of glucocorticoid receptor-associated hsp90 and glucocorticoid binding capacity of immunopurified untransformed receptors. Two conclusions are clear: 1) under no conditions is there any steroid binding capacity in the absence of receptor-associated hsp90; 2) there is a direct correlation between the amount of hsp90 bound to the receptor and the steroid binding capacity. These observations have led us to the conclusion that the binding of hsp90 to its binding site is necessary to generate the high affinity conformation of the steroid binding site [although binding of hsp90 is not in itself sufficient, as thiol reduction (Grippo et al., 1985) and perhaps other binding factors are also required]. Thus, the free energy involved in the binding of hsp90 to the hormone-free receptor may be transduced through a change in the conformation of the conserved "transducing" domain to yield a substantial decrease in the dissociation rate constant for the glucocorticoid. This model is particularly interesting from the general point of view of understanding how the binding of a drug or hormone is linked to the initial event in signal transduction. Indeed, it is analogous to experience with β-adrenergic receptors, where interactions between the receptor and the G protein subunit affect ligand binding affinity and vice versa. It is likely that the concept of a "transducing domain" will hold true for a variety of drug-receptor interactions.

A ROLE FOR hsp90 IN NUCLEAR TRANSPORT OF RECEPTOR

One requirement that is not accounted for in models of receptor structure that have been considered to date is that the receptor must make its way to its nuclear site of action by an organized process responsible for protein transfer to the nucleus, passage through the nuclear pores, and subsequent transfer to precise intranuclear sites of action. The sizes of glucocorticoid and progesterone receptors are considerably larger than the upper limits of size estimates for passage of molecules into the nucleus by diffusion (e.g., Lang et al., 1986) and it would seem that they have to be transported through the nuclear pore system. As estrogen and progesterone receptors are predominantly located in the nucleus prior to exposure to hormone, the receptors must go into some sort of "docking" mode once they are inside the nuclear envelope. It seems that the receptors are associated with hsp90, both during cytoplasmic-to-nuclear transfer and while they are "docked" in an untransformed state after arrival. This is inferred from the fact that the cytoplasmic glucocorticoid receptor and the nuclear estrogen and progesterone receptors are all recovered as 9S, receptor-hsp90 complexes after rupture of cells or tissues that have not been exposed to hormone. It is likely that the different relative abundance of the different steroid receptors in the cytoplasm vs. the nucleus, simply reflects different kinetics of overall receptor cycling into and out of the nucleus via the same system. Recent studies on the localization of both hsp90 and glucocorticoid receptor in intact cells may offer clues as to the nature of the structure responsible for receptor transport from cytoplasm to nucleus.

The avian virus transforming protein pp60^{src} is another protein that associates with hsp90 in intact cells, and it is thought that hsp90 is involved in the transport of pp60^{src} to the plasma membrane (Brugge, 1986). As is the case for the structure responsible for transport of proteins from cytoplasm to the nucleus, the structure responsible for transport of pp60^{src} to the plasma membrane has not been defined.

As hsp90 is an ubiquitous, abundant and highly conserved protein, we speculated that it may interact with another abundant and conserved protein in performing an essential cellular function related to protein transport. We therefore screened a number of monoclonal antibodies against structural proteins for their ability to cause the coimmunoadsorption of hsp90 from cytosols. We found that antibodies directed against tubulin could cause the coimmunoadsorption of hsp90 and then demonstrated by indirect immunofluorescence using the AC88 antibody against hsp90 that the heat shock protein is associated with microtubules in both interphase and mitotic marsupial epithelial cells (Sanchez et al., 1988). The immunofluoresence produced by AC88 is blocked by preabsorption with purified hsp90 but not by preabsorption with tubulin, showing that the fluorescence signal is specific for the localization of hsp90 and does not result from cross-reaction with tubulin (Welsh et al., submitted manuscript). Several mouse, rat and primate (including human) cell lines were examined, and in all cases, hsp90 distributed with microtubule structures.

It is most important to note that the immunofluorescence localization of hsp90 on tubulin-containing filaments in interphase cells is identical to the immunofluorescence localization of the glucocorticoid receptor in Rueber hepatoma and human uterine carcinoma cells reported by Wikstrom et al. (1987) using a monoclonal antibody specific for the receptor. We have noted that in interphase cells, the fibers defined by hsp90 immunofluorescence radiate from a focal point near the nucleus through all regions of the cytoplasm to the plasma membrane at the cell periphery (Welsh et al., submitted manuscript). These observations have led us to speculate (Sanchez et al., 1988) that the hsp90-tubulin-containing structures could provide a mechanism for transferring molecules from their cytoplasmic site of synthesis to the nucleus (e.g., steroid receptors) or to the plasma membrane (e.g., pp60^{src}). A type of microtubule-based movement has been visualized in neuraxons and in neuronal cell extracts (Vale et al., 1985a, 1985b) that might form a basis for conceptualizing such a mechanism of directed movement in the cell.

To ask whether any association between the glucocorticoid receptor and microtubules could be demonstrated under cell-free conditions, we heated L-cell cytosol at 37°C under tubulin polymerizing conditions (Pratt et al., 1989). We found that the majority of the tubulin and almost all of the receptor (which surprisingly is not cleaved) are converted to a form that pellets at 100,000 x g. The conversion of both the receptor and the tubulin to a particulate form is temperature-dependent and is not blocked by colchicine. When molybdate is present during heating of the cytosol, tubulin conversion to a particulate form is unaffected but the receptor remains entirely in the soluble form. If the conversion of receptor from a soluble to a particulate form is in any way related to the process of receptor translocation in intact cells, then specific features determined by nuclear localization signals should be required. Appropriate transfection experiments are being carried out with cDNAs determining various β-galactosidase-glucocorticoid receptor fusion proteins (Picard and Yamamoto, 1987) to see if biologically meaningful correlations can be established.

PREDICTIONS AND SPECULATIONS ARISING FROM A MODEL OF RECEPTOR TRANSPORT BY AN hsp90-MICROTUBULE SYSTEM

As indicated above, if a receptor is transported from the cytoplasm to the nucleus, it must contain signals that permit interaction with the transport mechanism. Two nuclear localization signals have been identified by Picard and Yamamoto (1987) in the primary sequence of the rat glucocorticoid receptor. As shown in Figure 2, one nuclear localization feature, NL1, maps to a 28 amino acid segment near the DNA-binding domain. This segment contains a short region of homology with the nuclear localization signal of SV40-T antigen. This sequence contains lys505 which Carlstedt-Duke et al., (1987) have unequivocally identified as the unique tryptic cleavage site that is hydrolyzed to generate the 27-kDa steroid-binding and 15-kDa DNA-binding

fragments of the glucocorticoid receptor. Thus, the NL1
site must lie on the surface of the receptor. The second
nuclear localization feature is located between position
532 and the carboxy terminus. NL2 confers glucocorticoid-
dependent nuclear localization on the receptor and on
fusion proteins.

It seems likely that NL1 of the glucocorticoid
receptor or a similar feature in the estrogen and
progesterone receptors is responsible for steroid-
independent transfer from the cytoplasm to "docking" sites
in the nucleus. The function of the second localization
signal, NL2, may be repressed by hsp90. Steroid-induced
dissociation of hsp90 from the receptor would then allow
progression from the hsp90-tubulin-containing "docking"
complex to a specific high affinity site of receptor
interaction in the nucleus.

Several possibilities derive from this model. First,
transport in this system may be bidirectional in the same
way that the axonal transport system can move both
antegrade and retrograde. Hsp90 may play a role in both
processes. For example, it is possible that as $pp60^{src}$ is
being synthesized, it binds to hsp90. Concomitant with or
subsequent to the binding of hsp90, a plasma membrane
localization feature on the viral protein that is analogous
to NL1 on the steroid receptor then binds the complex to an
antegrade system progressing in the direction of the plasma
membrane. It is known that different axoplamic proteins
are responsible for generating movement in opposite
directions along a microtubule scaffold (Vale et al.,
1985b), and it is likely that there are recognition signals
on transported proteins that are specific for the
directionality.

It is possible that the same outward moving
microtubule-based system that carries $pp60^{src}$ to the plasma
membrane also carries ribonucleoprotein particles from the
nucleus into the cytoplasm and to the endoplasmic
reticulum. The converse proposal is that a retrograde
system proceeding in the direction of the nucleus carries a
wide variety of large proteins having a nuclear site of
action. We have noted that the hsp90-containing filaments
identified by immunofluoresence form a continuous network
extending between the cell periphery and a perinuclear
dense fibrillar structure (Welsh et al., submitted

manuscript). This raises the possibility of direct transport of molecules from the inside of the plasma membrane to the nucleus. This possibility may be important when one considers that certain hormones and growth factors that bind to plasma membrane receptors have profound regulatory effects on the expression of specific genes in the cell nucleus. A protein that is phosphorylated at the plasma membrane by the insulin receptor, for example, could be "activated" for nuclear transport and proceed directly through the cytoplasm to its site of action on insulin-regulated genes in the nucleus in a manner analogous to the transport and nuclear delivery of steroid receptors.

The general role played by hsp90 in a prospective protein transport system is unknown. However, there are certain parallels that can be drawn from observations made in the pp60^{src} and the steroid receptor systems that lead one to speculate that at least one function of hsp90 is to regulate the ability of proteins to be transferred from the transport system. There are temperature-sensitive mutants of pp60^{src}, for example, that bind hsp90 with high affinity at nonpermissive temperature, a condition under which the viral protein does not become associated with the plasma membrane (Brugge et al., 1983). Thus, it may be that hsp90 must dissociate from pp60^{src} in order to expose some recognition feature on the protein that permits it to progress from the microtubule-based transport system to a site of interaction with the plasma membrane. In the case of steroid-mediated receptor transformation in the intact cell, ligand-mediated dissociation of hsp90 exposes a feature, perhaps NL2, that allows progression to high affinity nuclear "acceptor" sites. In the case of receptors that for some reason are located predominantly in the cytoplasmic portion of the transport system (e.g., glucocorticoid receptors), ligand-mediated dissociation of hsp90 may permit rapid progression through the system without a requirement for "docking" after passage through the nuclear pores. Thus, hsp90 may participate in protein transport proceeding in either direction and, at least with selected transported proteins, its presence may repress a feature of the protein that permits its progression from the transport system.

In their current state, these speculations are obviously primitive, but they permit the development of working models. In the model of the untransformed cytosolic receptor complex presented in Figure 2, hsp90 is shown to be repressing the function of the DNA-binding domain. This arrangement may be important with regard to understanding transformation of cytosolic receptors as a hormone-mediated derepression of the DNA-binding function. However, in the intact cell where we have defined transformation as the progression to a high affinity nuclear binding state, it is likely that it is the derepression of the NL2 function that is the immediate important consequence of hormone-mediated hsp90 dissociation.

REFERENCES

Bresnick EH, Dalman FC, Sanchez ER, Pratt WB (1989). Evidence that the 90-kDa heat shock protein is necessary but not sufficient for stabilizing the steroid binding state of the L-cell glucocorticoid receptor. Submitted manuscript.

Brugge JS (1986). Interaction of Rous sarcoma virus protein pp60^{src} with cellular proteins pp50 and pp90. Curr Topics Microbiol Immunol 123:1-22.

Brugge JS, Yonemoto W, Darrow D (1983). Interaction between the Rous sarcoma virus transforming protein and two cellular phosphoproteins: analysis of the turnover and distribution of this complex. Mol Cell Biol 3:9-19.

Carlstedt-Duke J, Stromstedt PE, Persson B, Cederlund E, Gustafsson JA, Jornvall H (1988). Identification of hormone-interacting amino acid residues within the steroid-binding domain of the glucocorticoid receptor in relation to other steroid hormone receptors. J Biol Chem 263:6842-6846.

Carlstedt-Duke J, Stromstedt PE, Wrange O, Bergman T, Gustafsson JA, Jornvall H (1987). Domain structure of the glucocorticoid receptor protein. Proc Natl Acad Sci USA 84:4437-4440.

Dalman FC, Sanchez ER, Lin ALY, Perini F, Pratt WB (1988). Localization of phosphorylation sites with respect to the functional domains of the mouse L-cell glucocorticoid receptor. J Biol Chem, in press.

Danielsen M, Northrop JP, Ringold GM (1986). The mouse glucocorticoid receptor: mapping of functional domains by cloning, sequencing and expression of wild-type and mutant receptor proteins. EMBO J 5:2513-2522.

Denis M, Gustafsson JA, Wikstrom AC (1988). Colocalization

of the M_r 90,000 heat shock protein with the steroid-binding domain of the glucocorticoid receptor. J Biol Chem, in press.

Grippo, JF, Holmgren A, Pratt WB (1985). Proof that the endogenous heat-stable glucocorticoid receptor-activating factor is thioredoxin. J Biol Chem 260:93-97.

Hollenberg SM, Giguere V, Segui P, Evans RM (1987). Colocalization of DNA-binding and transcriptional activation functions of the human glucocorticoid receptor. Cell 49:39-46.

Lang I, Scholz M, Peters R (1986). Molecular mobility and nucleocytoplasmic flux in hepatoma cells. J Cell Biol 102:1183-1190.

Mendel DB, Orti E (1988). Isoform composition and stoichiometry of the 90-kDa heat shock protein associated with glucocorticoid receptors. J Biol Chem 263:6695-6702.

Meshinchi S, Grippo JF, Sanchez ER, Bresnick EH, Pratt WB. Evidence that the endogenous heat-stable glucocorticoid receptor stabilizing factor is a metal component of the untransformed receptor complex. Submitted manuscript.

Picard D, Yamamoto KR (1987). Two signals mediate hormone-dependent nuclear localization of the glucocorticoid receptor. EMBO J 6:3333-3340.

Pratt WB (1987). Transformation of glucocorticoid and progesterone receptors to the DNA-binding state. J Cellular Biochem 35:51-68.

Pratt WB, Jolly DJ, Pratt DV, Hollenberg SM, Giguere V, Cadepond FM, Schweizer-Groyer G, Catelli MG, Evans RM, Baulieu EE (1988). A region in the steroid binding domain determines formation of the non-DNA-binding, 9S glucocorticoid receptor complex. J Biol Chem 263:267-273.

Pratt WB, Sanchez ER, Bresnick EH, Meshinchi S, Scherrer LC, Dalman FC, Welsh MJ (1989). Interaction of the glucocorticoid receptor with the 90-kDa heat shock protein: an evolving model of ligand-mediated receptor transformation and translocation. Cancer Research, in press.

Rexin M, Busch W, Gehring U (1988). Chemical cross-linking of heteromeric glucocorticoid receptors. Biochemistry 27:5593-5601.

Rusconi S, Yamamoto KR (1987). Functional dissection of the hormone and DNA-binding activities of the glucocorticoid receptor. EMBO J 6:1309-1315.

Sanchez ER, Meshinchi S, Tienrungroj W, Schlesinger MJ,

Toft DO, Pratt WB (1987). Relationship of the 90-kDa murine heat shock protein to the untransformed and transformed states of the L cell glucocorticoid receptor. J Biol Chem 262: 6986-6991.

Sanchez ER, Redmond T, Scherrer LC, Bresnick EH, Welsh MJ, Pratt WB (1988). Evidence that the 90-kDa heat shock protein is associated with tubulin-containing complexes in L-cell cytosol and in intact PtK cells. Mol Endocrinol 2:756-760.

Sanchez ER, Toft DO, Schlesinger MJ, Pratt WB (1985). Evidence that the 90-kDa phosphoprotein associated with the untransformed L-cell glucocorticoid receptor is a murine heat shock protein. J Biol Chem 260:12398-12401.

Simons SS, Pumphrey JG, Rudikoff S, Eisen HJ (1987). Identification of cysteine 656 as the amino acid of hepatoma tissue culture cell glucocorticoid receptor that is covalently labeled by dexamethasone 21-mesylate. J Biol Chem 262:9676-9680.

Smith LI, Bodwell JE, Mendel DB, Ciardelli T, North WG, Munck A (1988). Identification of cysteine-644 as the covalent site of attachment of dexamethasone 21-mesylate to murine glucocorticoid receptors in WEHI-7 cells. Biochemistry 27:3747-3753.

Vale RD, Schnapp BJ, Reese TS, Sheetz MP (1985a). Organelle, bead, and microtubule translocations promoted by soluble factors from the giant squid axon. Cell 40:559-569.

Vale RD, Schnapp BJ, Mitchison T, Steuer E, Rees TS, Sheetz MP (1985b). Different axoplasmic proteins generate movement in opposite directions along microtubules in vitro. Cell 43:623-632.

Welsh MJ, Sanchez ER, Bresnick EH, Redmond T, Schlesinger MJ, Toft DO, Pratt WB. Immunofluorescence localization of the 90-kDa heat shock protein to tubulin-containing structures in interphase and mitotic cells. Submitted manuscript.

Wikstrom AC, Baake O, Okret S, Bronnegard M, Gustafsson JA (1987). Intracellular localization of the glucocorticoid receptor: evidence for cytoplasmic and nuclear localization. Endocrinology 120:1232-1242.

Molecular Endocrinology and Steroid
Hormone Action, pages 133–155
© 1990 Alan R. Liss, Inc.

PHOSPHORYLATION OF UTERUS ESTRADIOL RECEPTOR ON TYROSINE

F. Auricchio, A. Migliaccio, G. Castoria,
M. Di Domenico and M. Pagano
Il Cattedra di Patologia Generale
Istituto di Patologia Generale, I Facolta di
Medicina e Chirurgia,
Universita di Napoli, Italy

INTRODUCTION

The possibility that phosphorylation of proteins is responsible for the hormone binding to steroid receptors was initially suggested for the glucocorticoid receptor when it was observed that ATP shortage in thymocytes decreased their ability to bind hormone whereas ATP recovery paralleled recovery of this binding (Munck and Brinck-Johnsen, 1968). Subsequently, it was observed that hormone binding to different steroid receptors is inactivated by phosphatases and reactivated by processes requiring ATP (Sando et al., 1979a, 1979b; Liao et al., 1980).

We partially purified two uterus enzymes that subsequently were identified as a phosphatase and a kinase (Auricchio and Migliaccio, 1980; Auricchio et al., 1981a, 1981b). The phosphatase inactivates the hormone binding of the uterus estradiol receptor (Auricchio and Migliaccio, 1980; Auricchio et al., 1981a), the kinase reactivates the phosphatase-inactivated binding (Auricchio et al., 1981b). Since most of the estradiol receptor in the uterus is phosphorylated and binds hormone, preincubation of the receptor with the phosphatase was required to demonstrate in cell-free system that estradiol receptor binding requires phosphorylation on tyrosine of the receptor (Migliaccio et al., 1982; Migliaccio et al., 1984). Phosphorylation of proteins on tyrosine is about a thousand fold less frequent than phosphorylation on serine and threonine (Hunter and Cooper, 1985) and seems to be involved in important processes like growth factor induced cell multiplication, cell transformation and cell differentiation (Bishop, 1985; Hunter and Sefton, 1980; Hafen et al., 1987). We will briefly review some properties of the two enzymes regulating the hormone binding of the estradiol receptor and the evidence that the estradiol receptor is phosphorylated on tyrosine in whole uterus (Migliaccio et al., 1986).

REGULATION OF UTERUS ESTRADIOL RECEPTOR-TYROSINE KINASE BY Ca^{2+}-CALMODULIN AND ESTRADIOL

This enzyme converts the non-hormone binding into hormone binding receptor through phosphorylation of the receptor on tyrosine (Migliaccio et al., 1984). It can be purified from receptor-rich cytosol of 30-50g calf uterus and is routinely assayed by its ability to reactivate the binding or to rephosphorylate the calf uterus estradiol receptor inactivated and dephosphorylated by the calf uterus nuclear phosphatase (Auricchio et al., 1987a). This enzyme activity is unstable. The Michaelis constant for the dephosphorylated receptor in optimal conditions of assay is 0.3 nM (Auricchio et al., 1981b). This high affinity is strong evidence that non-phosphorylated, non-hormone binding receptor is a natural substrate of this kinase.

Regulation of the activity of this enzyme in cell-free systems is rather complex. In fact the kinase is stimulated by Ca^{2+}-calmodulin (Migliaccio et al., 1984) as well as by estradiol-receptor complex (Auricchio et al., 1987b). Using the hormone-binding activation assay performed with crude substrate, it was initially observed that Ca^{2+} stimulates the kinase activity (Auricchio et al., 1981b). To assess whether Ca^{2+} stimulation is mediated by calmodulin, homogeneous receptor which was calmodulin-free and partially inactivated by the purified nuclear receptor-phosphatase was used as a substrate in the hormone-binding activation assay (Migliaccio et al., 1984). It was found that combined Ca^{2+} and calmodulin is required for kinase stimulation by Ca^{2+} (Migliaccio et al.,1984). Alone, neither Ca^{2+} nor calmodulin produces a stimulatory effect. Dose-response curves for calmodulin and Ca^{2+} stimulation of hormone-binding activation by the kinase have been calculated. The half-maximal and maximal rates of activation are reached at approximately 60 and 600 nM calmodulin and 0.8 and 1 µM Ca^{2+}, respectively (Migliaccio et al., 1984). The high affinity of the kinase for calmodulin prompted us to use calmodulin-Sepharose to purify the kinase further (Auricchio et al., 1987a). Ca^{2+}-calmodulin stimulates binding activation as well as phosphorylation on tyrosine of the estradiol receptor in parallel fashion confirming that phosphorylation on tyrosine is required to bind hormone to the receptor (Migliaccio et al., 1984).

As regard to stimulation of the kinase by estradiol, in a preliminary experiment on a crude system it was observed that the ability of the kinase to activate estradiol specific binding sites of phosphatase-inactivated receptor in the presence of phosphorylated receptor, barely detectable in the absence of exogenous estradiol, is drastically stimulated by receptor pre-incubated with exogenous hormone (Auricchio et al., 1986). Since activation of binding sites is linked with receptor phosphorylation it was expected that estradiol also stimulates phosphorylation of its own receptor. Recent experiments including those reported in the following paragraph prove this point.

The receptor, purified from calf uterus by ammonium sulphate precipitation and heparin-Sepharose chromatography, was partially inactivated by the nuclear phosphatase and preincubated in the absence or in the presence of 4 nM [^{3}H]estradiol followed by incubation with partially purified kinase and [γ-^{32}P]ATP. Control receptor-less or kinase-less samples preincubated with hormone were also run. After incubation with monoclonal antibody against estradiol receptor (Moncharmont et al., 1982) all the samples were treated with protein A-Sepharose (Pansorbin) to precipitate the antibodies and proteins associated with antibodies. The proteins eluted from the pellets were submitted to SDS-PAGE followed by autoradiography (Fig. 1), and in a different experiment, to phosphoaminoacid analysis (panel A of Fig. 2).

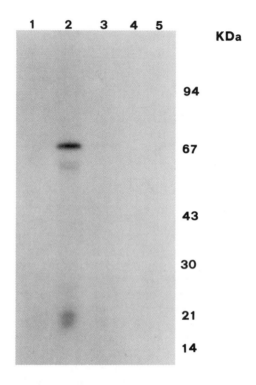

Fig. 1. SDS-polyacrylamide gel electrophoresis of the immunoprecipitated estradiol receptor phosphorylated in the absence and in presence of hormone. Receptor partially purified from calf uterus was inactivated by the phosphatase obtained by CM-cellulose chromatography of the calf uterus nuclear extract (20% of the binding sites were inactivated and preincubated at 0° C for 3 h. in the absence (lane 3) and in presence of 4 nM ^{3}H estradiol (lanes 1 and 2, respectively). Samples were then incubated with the partially purified kinase from calf uterus at 15° C in TGD-buffer (Tris-HCl 50 mM, EGTA 0.2 mM, DTT 1mM; pH 7.4) containing 5 mM $MgCl_2$, 10 mM sodium molybdate, 0.8 mM $CaCl_2$, 10 µg/ml calmodulin and 0.15 mM [γ-^{32}P]ATP (10 Ci/mmol). Receptor-less and kinase-less controls were run in parallel after preincubation at 0° C for 3 h. in the presence of 4 nM [^{3}H]estradiol (lane 5 and lane 4, respectively).

(Fig 1. Continued)
At the end of the incubation, to each sample was added either JS 34/32 antireceptor (lanes 2-5) or control antibodies (lane 1) and Pansorbin suspension, then centrifuged. Aliquots of proteins eluted with SDS-PAGE sample buffer from the pellets were submitted to SDS-PAGE and the gel dried and autoradiographed. Activation of estradiol specific binding sites by the kinase in the absence and presence of hormone was measured in parallel samples (0.011 pmoles and 0.107 pmol, respectively, the latter value corresponding to a complete reactivation of the phosphatase-inactivated binding sites). (Auricchio et al., 1987b).
In the incubation mixture containing estradiol receptor incubated with hormone and kinase (lane 2 of Fig. 1) the receptor has been phosphorylated. In fact, autoradiography of this lane showed a more phosphorylated band migrating as a 68 Kd protein and lighter and less phosphorylated proteins, probably proteolytic products of the 68 Kd receptor (Migliaccio et al., 1986; Van Osbree et al., 1984; Katzenellenbogen et al., 1983; Walter et al., 1985). In contrast, when the same incubation was performed with receptor preincubated without estradiol, phosphorylation of the receptor was barely detectable (lane 3 of Fig. 1). Comparison of lane 2 with lane 3 and assay of activation of binding sites by the kinase in the presence and absence of hormone in parallel samples (0.107 and 0.011 pmol, respectively) showed that estradiol strongly stimulates both phosphorylation of receptor and activation of hormone binding. No phosphorylation was detected by preincubation of the receptor with hormone followed by incubation in the absence of kinase (lane 4). This confirmed our previous observation made under different conditions that the receptor has no kinase activity (Migliaccio et al., 1982). Almost no detectable phosphorylation of the 68 Kd band of the receptor was present in the sample containing kinase preincubated with hormone and incubated without adding receptor (lane 5). This very faint phosphorylation was not observed with other kinase preparations and was due to co-purification with the partially purified kinase of small amounts of dephosphorylated as well as hormone binding receptor from calf uterus cytosol (Auricchio et al., 1987b). Lane 1 shows a sample identical to that shown by lane 2 except for incubation with N6 control antibody. No ^{32}P-phosphorylated protein was immunopre-

cipitated in this sample proving the specificity of the receptor immunoprecipitation by JS 34/32 antibody. In conclusion, this experiment shows that estradiol strongly stimulates both phosphorylation and reactivation of hormone binding of the phosphatase-inactivated receptor by the kinase.

Several findings prove that estradiol stimulation of the kinase on the phosphatase-inactivated receptor requires the hormone-receptor complex; tamoxifen inhibited to a similar extent hormone binding to the receptor and the stimulatory effect of estradiol on the kinase; steroid hormones different from estradiol did not stimulate the kinase; maximal stimulation of the kinase was observed at physiological concentration of hormone (Auricchio et al., 1987b). In addition the possibility that estradiol directly stimulates the kinase was excluded by the lack of estradiol binding to the kinase extensively purified by calmodulin-Sepharose at physiological concentrations of hormone (Auricchio et al., 1987b). Kinase stimulation by estradiol and Ca^{2+}-calmodulin has been confirmed using exogenous substrates of this enzyme as well as in vitro synthesized human estradiol receptor (manuscripts in preparation). Phosphoaminoacid analysis of the receptor which was ^{32}P-phosphorylated and reactivated by the kinase in the absence and in the presence of estradiol has been performed either on the receptor immunoprecipitated (panel A of Fig. 2), or on the 68 Kd protein band eluted from the SDS-PAGE of the immunoprecipitated receptor (panel B of Fig. 2).

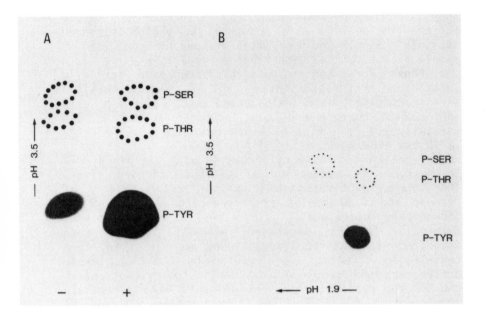

Fig. 2. **Phosphoaminoacid analysis of the estradiol receptor phosphorylated by the kinase in the absence and in the presence of hormone.** Panel A: samples containing calf uterus receptor partially inactivated by the phosphatase (30% of the binding sites were inactivated) were preincubated at 0° C for 3 h. in the absence and presence of hormone (4 nM [³H]estradiol) and incubated at 15° C with the kinase and [γ-³²P]ATP, as described in the legend to Fig. 1. Parallel samples were incubated in the same conditions in the absence and presence of radioinert ATP to measure the activation of estrogen binding sites by the kinase. In the absence of hormone, only 0.039 pmoles were activated, but 0.145 pmoles, corresponding to 97% of the phosphatase-inactivated binding sites were activated in presence of hormone. Samples incubated with

(Fig. 2, Continued)

[ɤ-³²P]ATP were mixed with 15 µl rat ascites containing JS 34/32 anti-receptor antibody and 10 µl 1% Pansorbin, then centrifuged. Samples eluted with SDS-PAGE sample buffer from pellets were submitted to phosphoaminoacid analysis. Sample phosphorylated without hormone: lane –; sample phosphorylated with hormone: lane +.

Panel B: samples containing receptor partially inactivated by the phosphatase (36% of the binding sites were inactivated) were preincubated at 0° C for 3 h. with and without hormone and then incubated at 15° C with partially purified kinase in the presence of [ɤ-³²P]ATP as in the experiment of panel A.

Parallel samples were incubated with radioinert ATP to measure the activation of estrogen binding sites. The kinase reactivated 0.01 pmoles in the absence of hormone and 0.30 pmoles (corresponding to 97% of the phosphatase-inactivated binding sites) in the presence of hormone. Samples incubated with [ɤ-³²P]ATP were incubated with antireceptor antibody and Pansorbin, then centrifuged. Pellets were eluted by SDS-PAGE sample buffer, and each sample was divided in two aliquots of 15 and 55 µl, respectively. The aliquots were separately submitted to SDS-PAGE. The lanes loaded with 15 µl aliquots were dried and submitted to autoradiography. No phosphorylated receptor was detected in the sample incubated in absence of hormone and therefore no phosphoaminoacid analysis was performed on this sample. The 68 Kd phosphorylated protein was extracted from the SDS-PAGE loaded with the 55 µl aliquot of the sample incubated in the presence of hormone, hydrolyzed, lyophilized and submitted to phosphoaminoacid analysis (Auricchio et al., 1987b). In the first case (panel A) the phosphoaminoacid electrophoresis was run at pH 3.5 in one direction. In the second case (panel B) it was run at pH 1.9 in the first direction, and at pH 3.5 in the second direction. In both cases phosphoaminoacid analysis showed that the receptor has been phosphorylated on tyrosine confirming a previous report on the ability of the uterus kinase to phosphorylate the receptor on tyrosine (Migliaccio et al., 1984). Fig. 2 shows that estradiol stimulates phosphorylation of the receptor on tyrosine. In the experiment of panel A, phosphorylation (and activation of hormone binding) was observed, although at a lower level, also in the absence of exogenous hormone. Conversely, in a different

(Fig. 2, Continued)

experiment, in the absence of estradiol (panel B) no phosphorylation of the immunoprecipitated receptor submitted to SDS-PAGE (as well as no significant reactivation of binding) was detectable. Therefore the experiments in Fig. 2 confirm the association of hormone binding reactivation and phosphorylation on tyrosine of receptor by the kinase. Actually, the extent of stimulation of the kinase by estradiol is different in the different experiments and this might be due to different amounts of endogenous estradiol complexed with the purified receptor preparations.

Regulation of the estradiol receptor tyrosine kinase shows some analogy with that of the insulin receptor-associated tyrosine kinase since both tyrosine kinases are stimulated by Ca^{2+}-calmodulin (Graves et al., 1985; Graves et al., 1986) as well as by hormone occupancy of the corresponding receptor (Kasuga et al., 1982). Hormone occupancy of receptors also stimulates other receptor-associated tyrosine kinases like the EGF, PDGF and somatomedin C receptor-associated tyrosine kinases (Cohen et al., 1980; Ek and Heldine, 1982; Jacobs et al., 1983).

Table 1 summarizes some of the properties of the estradiol receptor kinase.

TABLE 1. Properties of the estradiol receptor-kinase purified from calf uterus

---

1) It phosphorylates the receptor exclusively on tyrosine converting the non hormone-binding into hormone-binding receptor.
2) It is stimulated by Ca^{2+}-calmodulin.
3) It is stimulated by estradiol-receptor complex. K_m for the non phosphorylated receptor: 0.3 nM.

THE PHOSPHATASE INACTIVATING THE HORMONE BINDING OF THE RECEPTOR IS AN ESTRADIOL RECEPTOR-PHOSPHOTYROSINE PHOSPHATASE.

The phosphatase has been found in the nuclei of mouse mammary gland and calf and mouse uterus (Auricchio and Migliaccio 1980). It is not present in mouse quadriceps muscle nuclei (Auricchio and Migliaccio 1980). It is completely inhibited by several phosphatase inhibitors including protein-phosphotyrosine phosphatase inhibitors like zinc and vanadate (Auricchio and Migliaccio 1980; Migliaccio et al., 1986). In vitro the enzyme inactivates the hormone binding of crude and pure cytosol receptor (Auricchio et al., 1981a). In vitro inactivation, apparently due to dephosphorylation of estradiol receptor, has been observed after receptor "translocation" into nuclei of mouse uterus injected with estradiol and is attributed to this phosphatase (Auricchio et al., 1982). This enzyme inactivates hormone-free as well as hormone-bound receptor (Auricchio et al., 1981a).

The very high affinity of the hormone binding inactivating activity of the phosphatase for the receptor ($K_m \sim 1$ nM, Auricchio et al., 1981a) lends weight to our hypothesis that this receptor is a physiological substrate of this enzyme.

TABLE 2. Properties of the estradiol receptor phosphatase.

----

1) Localized in nuclei of estrogen target tissues.
2) Purified from calf uterus.
3) Inhibited by different phosphatase inhibitors like zinc, molybdate, fluoride, phosphate, pyrophosphate, p-nitrophenyl phosphate and orthovanadate.
4) It inactivates the hormone binding of cytosol and nuclear receptor.
5) Km for estrogen-free receptor: 1.5 nM.
6) Km for estrogen-bound receptor: 0.8 nM.
7) It does not inactivate the hormone binding of the receptor complexed with non-steroidal antiestrogens.
8) It dephosphorylates the estradiol receptor phosphorylated by the receptor tyrosine-kinase.
9) It abolishes the interaction of estradiol receptor with the 2G8 antiphosphotyrosine antibody.

That the phosphatase inactivating the hormone binding of the estradiol receptor dephosphorylates phosphotyrosyl residue(s) of the receptor is shown by the following two findings. Phosphorylation by the kinase of phosphatase-inactivated receptor using $[\gamma-^{32}P]ATP$ produces reactivated receptor which is ^{32}P-phosphorylated exclusively on tyrosine (Migliaccio et al., 1984). Incubation of this ^{32}P-receptor in the presence of the phosphatase partially inactivates the hormone binding and removes significant amount of ^{32}P incorporated into the receptor as shown by SDS-PAGE presented in Fig. 3 (Auricchio et al., 1984).

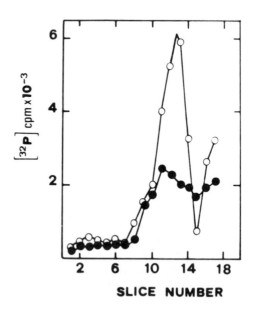

Fig. 3. SDS-polyacrylamide gel electrophoresis of the phosphorylated receptor before and after incubation with the phosphatase. Highly purified estradiol-17β receptor was partially inactivated by incubation with the phosphatase purified by CM-cellulose chromatography of the nuclear extract from calf uterus and used as substrate of the partially purified kinase in the presence of 0.15 mM [γ-³²P]ATP under conditions similar to those described in the legend to Fig. 1. Two sample aliquots were incubated in the absence and presence of the phosphatase respectively then submitted to SDS-gel electrophoresis. Then the gel lanes were sliced and counted for ³²P radioactivity. Open symbols: receptor incubated in the absence of the phosphatase. Closed symbols: receptor incubated in the presence of the phosphatase (Auricchio et al., 1984).

Purified calf uterus estradiol receptor interacts with high affinity with 2G8 and 1G2 antiphosphotyrosine antibodies coupled to Sepharose beads (Kd 0.28 and 1.11 nM respectively; Fig. 4) and crude calf uterus receptor interacts with 2G8 antiphosphotyrosine antibodies (Auricchio et al., 1987c).

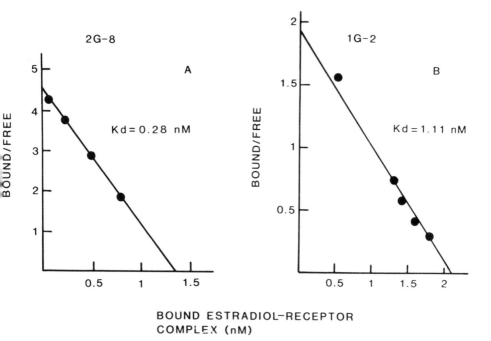

Fig. 4. **Measurement** of the affinity of the estradiol-receptor complex for 2G8 and 1G2 anti-phosphotyrosine antibodies coupled to Sepharose. Two samples of 1.5 ml of calf uterus cytosol [³H]estradiol-receptor complex preparation extensively purified according to the procedure previously reported (Van Osbree et al., 1984) were separately incubated with 0.1 vol of packed BSA-Sepharose beads for 2h at 0° C to

(Fig. 4, Continued)
remove molecules nonspecifically bound to protein-Sepharose. The suspensions were centrifuged and different aliquots of each supernatant containing 25-500 fmol (A) and 380-1500 (B) of [³H]estradiol-specific binding sites were diluted with TGD-buffer to 0.35 ml. (A) and 0.2 ml (B). Each aliquot was incubated under gentle shaking overnight at 0° C with 35 µl of packed 2G8 anti-P-tyr-Sepharose beads (panel A) and 20 µl of 1G2 anti-P-tyr-Sepharose (panel B), then centrifuged. The specific hormone-binding sites found in the supernatant after charcoal treatment were used as a measure of free [³H]hormone-receptor complex. Bound [³H]-hormone-receptor complex was calculated from the difference between specific binding sites present in samples before and after incubation with anti-P-tyr-Sepharose. B/F, bound/free hormone-receptor complex (Auricchio et al., 1987c).

Incubation of crude calf uterus cytosol receptor with homologous nuclei containing the receptor-phosphatase inactivates a portion (about 30%) of the receptor. This portion does not bind to 2G8 antiphosphotyrosine and is reactivated by incubation with ATP in the presence of the kinase (Migliaccio et al., 1986). This observation confirms that the phosphatase hydrolyses receptor-phosphotyrosine and inactivates the receptor (Auricchio et al., 1984). From this experiment it appears that phosphorylation on tyrosine of the receptor in calf uterus is required for hormone binding.

NON STEROIDAL ANTIESTROGENS INHIBIT INACTIVATION-REACTIVATION OF ESTRADIOL RECEPTOR IN CELL FREE-SYSTEMS.

In a previous report we observed that nonsteroidal antiestrogens like tamoxifen and nafoxidine once complexed either in vitro or in vivo with uterus estradiol receptor protect the hormone binding of the receptor from the in vitro inactivation by the nuclear phosphatase (Auricchio et al., 1981c). This finding, when related to the observation that non-steroidal antiestrogen-receptor complexes, in contrast with estrogen receptor complex, are slowly lost in nuclei of intact cells (Horwitz and McGuire, 1978), supports the possibility that the nuclear phosphatase is responsible for the loss of receptor "translocated" into nuclei by the hormone in vivo.

As previously reported in this chapter estradiol in complex with its hormone binding receptor stimulates the receptor-tyrosine kinase. Tamoxifen, at a concentration which reduces the occupancy of the receptor by estradiol by 50 per cent, also reduces the stimulatory effect of the hormone on the kinase by 45 per cent (Auricchio et al., 1987b). This experiment shows that tamoxifen by its interaction with the receptor prevents the stimulation of the kinase by estradiol.

In synthesis in cell-free systems, antiestrogens inhibit inactivation as well as activation of hormone binding of the receptor decreasing the reversible conversion between the hormone binding and non-hormone binding forms of receptor. Both effects are reached through the formation of the receptor-antihormone complex since this complex, unlike the receptor-hormone complex, is recognized neither as a substrate by the phosphatase nor as an activator by the kinase.

PHOSPHORYLATION OF ESTRADIOL RECEPTOR IN WHOLE UTERUS

To investigate whether estradiol receptor is phosphorylated on tyrosine not only in cell-free systems but also in whole tissues, uteri from intact young adult rats were incubated with ^{32}P-orthophosphate at 39°C for 1h 80 µM Na_3VO_4 was added to the incubation buffer. The uteri were mixed with carrier uteri and homogenized (Migliaccio et al., 1986) to prepare high speed supernatant. This supernatant was used to purify estradiol receptor by cycling it through diethyl-stilbestrol (DES) Sepharose column (Van Osbree et al., 1984). The receptor was eluted from the affinity resin by [^{3}H]estradiol, cycled through a heparin-Speharose column and finally eluted from this resin by a buffer containing heparin (Van Osbree et al., 1984). Estradiol receptor eluted from heparin-Sepharose was equilibrated with 12 nM [^{3}H]estradiol and incubated with an excess of immunoglobulins. These immunoglobulins were purified either from a control hybridoma derived from the fusion of myeloma cells with spleen cells from non-immunized mice or from the JS 34/32 clone produced by fusion of myeloma cells with spleen cells from mice immunized with purified estradiol receptor (Moncharmont et al., 1982). The two samples were analyzed by centrifugation through "high salt" sucrose gradients. The [^{3}H]estradiol peak bound to the receptor incubated with control immunoglobulins cosediments at 4.5 S with a peak of ^{32}P. Preincubation of the receptor preparation with JS 34/32 antibodies against purified receptor causes both peaks to shift to 7.5S. Since this shift is due to the formation of an antibody-receptor complex in a 1:1 molar ratio (Moncharmont et al., 1982) it is clear that the ^{32}P peak shifted to 7.5S by the antibodies belongs to the receptor (Migliaccio et al., 1986).

The receptor eluted from heparin-Sepharose was further purified by chromatography through a column of Sepharose to which JS 34/32 monoclonal antibodies against the receptor have been linked (Cuatrecasas, 1970). The receptor sample was eluted from the antibody-Sepharose column at alkaline pH, neutralized after the elution and concentrated by acid precipitation using myoglobin as a carrier. The pellet was dissolved and submitted to SDS-Polyacrylamide gel electrophoresis (Fig. 5).

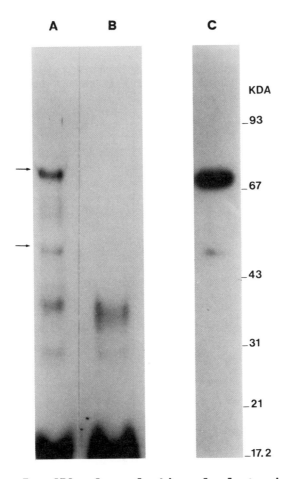

<u>**Fig. 5.**</u> **SDS-polyacrylamide gel electrophoresis of
the estradiol receptor purified from uteri of young adult
and intact rats incubated with ³²P-orthophosphate.**
Receptor was purified by DES-Sepharose, heparin-Sepharose
and antireceptor antibody-Sepharose chromatographies.
Receptor preparation was added with 10% TCA using
myoglobin as a carrier. The pellet was washed and
dissolved with SDS-PAGE sample buffer then heated at 100°
C for 3 min. An aliquot was submitted to SDS-PAGE.
After the run, the gel was stained with silver stain
(Bio-Rad), dried and exposed to autoradiography. Lane A:
silver staining of the receptor sample added with
myoglobin; Lane B: silver staining of myoglobin control

(Fig. 5, Continued)
sample; Lane C: autoradiography of lane A. The arrows show the two protein bands present only in the receptors sample (Migliaccio et al., 1986).

Silver nitrate stained several protein bands; only two of them (those indicated by arrows in Fig. 5) belong to the receptor preparation since they were not detectable in the control sample of myoglobin. The molecular weights of these two proteins were 68 and 48 Kd, respectively. Autoradiography showed a heavily phosphorylated band coincident with the 68 Kd protein and a slightly phosphorylated band coincident with the lighter protein that is probably a proteolytic product of the 68 Kd receptor (Migliaccio et al., 1986; Van Osbree et al., 1984; Katzenellenbogen et al., 1983; Walter et al., 1985).

A sample of the immunopurified receptor was subjected to acid hydrolysis, concentrated by lyophilization, dissolved in a small volume of water, and analyzed by one-dimensional electrophoresis at pH 3.5. The electrophoresis plate was exposed for autoradiography. The only phosphorylated aminoacid detectable was phosphotyrosine (Fig. 6).

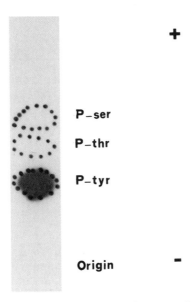

Fig. 6. **Phosphoaminoacid analysis of the estradiol receptor purified from rat uteri incubated with** [32]**P-orthophosphate.** Receptor purified according to the procedure reported in the legend to Fig. 5 was subjected to acid hydrolysis. An aliquot of this sample was submitted to phosphoaminoacid analysis by electrophoresis at pH 3.5. The plate was stained with ninhydrin and then exposed to autoradiography. The dotted lines represent the standard superimposed on autoradiography (Migliaccio et al., 1986).

This was the first demonstration of a steroid receptor phosphorylation on tyrosine in whole tissues. Recently the glucocorticoid receptor has been reported to be phosphorylated on both serine and tyrosine in human breast epithelial cells (Rao and Fox, 1987). The absence of a phosphoaminoacid different from phosphotyrosine in the phosphoaminoacid analysis of the whole rat uterus estradiol receptor does not exclude the possibility that other phosphoaminoacids are present in this receptor which might be detected under different experimental conditions (Migliaccio et al., 1986). Most of the proteins phosphorylated on tyrosine are also phosphorylated on different aminoacids.

ACKNOWLEDGMENTS

The authors gratefully acknowledge Mr. Gian Michele La Placa for editorial work. This research was supported by grants from Associazione Italiana per la Ricerca sul Cancro, from Italian National Research Council, Special Project Oncology, contract No. 87.1167.44 and from Ministero Pubblica Istruzione, Italy.

REFERENCES

Auricchio F., Migliaccio A. (1980). Inactivation of estradiol receptor by nuclei: prevention by phosphatase inhibitors. FEBS Lett. 117:224-226.

Auricchio F., Migliaccio A., Rotondi A. (1981a). Inactivation of estrogen receptor "in vitro" by nuclear dephosphorylation. Biochem. J. 194:569-574.

Auricchio F., Migliaccio A., Castoria G., Lastoria S., Schiavone E. (1981b). ATP-dependent enzyme activating hormone binding of estradiol receptor. Biochem. Biophys. Res. Commun. 101:1171-1178.

Auricchio F., Migliaccio A., Castoria G. (1981c). Dephosphorylation of estradiol nuclear receptor in vitro. A hypothesis on the mechanism of action of non-steroidal anti-estrogens. Biochem. J. 198:699-702.

Auricchio F., Migliaccio A., Castoria G., Lastoria S., Rotondi A. (1982). Evidence that "in vivo" estradiol receptor translocated into nuclei is dephosphorylated and released into cytoplasm. Biochem. Biophys. Res. Commun. 106:149-157.

Auricchio F., Migliaccio A., Castoria G., Rotondi A., Lastoria S. (1984). Direct evidence of "in vitro" phosphorylation-dephosphorylation of estradiol-17β receptor. Role of Ca²⁺-calmodulin in the activation of hormone binding sites. J. Steroid Biochem. 20:31-35.

Auricchio F., Migliaccio A., Castoria G., Rotondia A., Di Domenico M., Pagano M. (1986). Activation-inactivation of hormone binding sites of the estradiol 17β-receptor is a multiregulated process. J. Steroid Biochem. 24:39-43.

Auricchio F., Migliaccio A., Castoria G., Rotondi A., Di Domenico M. (1987a). Calmodulin-stimulated estradiol receptor-tyrosine kinase. In Means AR., Conn. MP (eds): "Methods in Enzymology" New York; Academic Press 139:731-744.

Auricchio F., Migliaccio A., Di Domenico M., Nola E. (1987b). Estradiol stimulates phosphorylation of its own receptor in a cell-free system. EMBO J. 6:2923-2929.

Auricchio F., Migliaccio A., Castoria G., Rotondi A., Di Domenico M., Pagano M., Nola E. (1987c). Phosphorylation on tyrosine of estradiol-17β receptor in uterus and interaction of estradiol-17β and glucocorticoid receptors with antiphosphotyrosine antibodies. J. Steroid Biochem. 27:254-253.

Bishop M. (1985) Viral oncogenes. Cell 42:23-38.

Cohen S., Carpenter G., King L. (1980). Epidermal growth-factor receptor protein kinase interactions: co-purification of receptor and epidermal growth factor enhanced phosphorylation. J. Biol. Chem. 255:4834.

Cuatrecasas P. (1970). Protein purification by affinity chromatography. Derivatization of agarose and polyacrylamide beads. J. Biol. Chem. 245:3059-3065.

Ek B., Heldin C.H. (1982). Characterization of a tyrosine specific kinase activity in human fibroblast membranes stimulated by platelet derived growth factor. J. Biol. CHem. 257:10486-10492.

Graves C.B., Goewert R.R., McDonald J.M. (1985). The insulin receptor contains a calmodulin-binding domain. Science 230:827-829.

Graves C.B., Gale R.D., Laurino J.P., McDonald J.M. (1986). The insulin receptor and calmodulin. Calmodulin enhance insulin-mediated receptor kinase activity and insulin stimulates phosphorylation of calmodulin. J. Biol. Chem. 261:10429.

Hafen E., Basler K., Edstrom J.E., Rubin G.M. (1987). Sevenless, a cell-specific homeotic gene of Drosophila, encodes a putative transmembrane receptor with a tyrosine kinase domain. Science 236:55-63.

Horwitz K.B., McGuire W.L. (1978). Nuclear mechanism of estrogen action: effect of estradiol and anti-estrogens on estrogen receptors and nuclear receptor processing. J. Biol. Chem. 23:8185-8191.

Hunter T., Sefton B. (1980). Transforming gene product of Rous sarcoma virus phosphorylates tyrosine. Proc. Natl. Acad. Sci. USA 77:1311-1315.

Hunter T., Cooper J.A. (1985). Protein tyrosine Kinase. Ann. Rev. Biochem. 54:897-930.

Jacobs S., Cull F.C., Earp H.S., Svoboda M.E., Van Wyk J.J., Cuatrecasas P. (1983). Somatomedin-C stimulates the phosphorylation of the B subunit of its own receptor. J. Biol. Chem 258:9581-9584.

Kasuga M., Karlsson F.A., Kahn C.R. (1982). Insulin stimulates the phosphorylation of the 95,000 dalton subunit of its own receptor. Science 215:185-187.

Katzenellenbogen J.A., Carlson K.E., Heiman D.F., Robertson D.W., Wei L.L., Katzenellenbogen B.S. (1983). Efficient and highly selective covalent binding labelling of the estrogen receptor with ³H-tamoxifen aziridine. J. Biol. Chem. 258:3487-3495.

Liao S., Rossini G.P., Hiipakka R.A., Chen C. (1980). Factors that can control the interaction of the androgen-receptor complex with the genomic structure in the rat prostate. In Bresciani F. (ed): "Perspectives in Steroid Receptor Research" New York: Raven Press, pp. 99-112.

Migliaccio A., Lastoria S., Moncharmont B., Rotondi A., Auricchio F. (1982). Phosphorylation of calf uterus 17β-estradiol receptor by endogenous Ca²⁺-stimulated kinase activating the hormone binding of the receptor. Biochem. Biophys. Res. Commun. 109:1002-1010.

Migliaccio A., Rotondi A., Auricchio F. (1984). Calmodulin stimulated phosphorylation of 17β-estradiol receptor on tyrosine. Proc. Natl. Acad. Sci. USA 81:5921-5925.

Migliaccio A., Rotondi A., Auricchio F. (1986). Estradiol receptor: phosphorylation on tyrosine in uterus and interaction with antiphosphotyrosine antibody. EMBO J. 5:2867-2872.

Moncharmont B., Su J.L., Parik I. (1982). Monoclonal antibodies against estrogen receptor: interaction with different molecular forms and functions of the receptor. Biochemistry 21:6916-6921.

Munck A., Brinck-Johnsen T. (1968). Specific and non-specific physiological interaction of glucocorticoids and related steroids with rat thymus cells in vitro. J. Biol. Chem. 243:5556-5565.

Roa K.V.S., Fox F. (1987). Epidermal growth factor stimulates tyrosine phosphorylation of human glucocorticoid receptor in cultured cells. Biochem. Biophys. Res. Commun. 144:512-519.

Sando J.J., Hammond N.D., Stratford C.A., Pratt W.B. (1979a). Activation of thymocyte glucocorticoid receptors to the steroid binding form. J. Biol. Chem. 254:4779-4789.

Sando J.J., LaForest A.C., Pratt W.B. (1979b). ATP-dependent activation of cell glucocorticoid receptors to the steroid binding form. J. Biol. Chem. 254:4772-4778.

Van Osbree T.R., Kim U.H., Mueller G.C. (1984). Affinity chromatography of estrogen receptors on diethyl-stilbestrol-agarose. Anal. Biochem. 136:321-327.

Walter P., Green S., Green G. (1985). Cloning of the human estrogen receptor cDNA. Proc. Natl. Acad. Sci. USA 82:7889-7893.

III. REGULATION OF GENE EXPRESSION IN STEROID DEPENDENT AND INDEPENDENT MODELS

Molecular Endocrinology and Steroid
Hormone Action, pages 159–169
© 1990 Alan R. Liss, Inc.

REGULATION OF PROLACTIN GENE EXPRESSION BY ESTRADIOL

Richard A. Maurer, Kyoon E. Kim, Richard N.
Day and Angelo C. Notides

Department of Physiology and Biophysics,
University of Iowa, Iowa City Iowa 52242
(R.A.M., K.E.K., R.N.D) and Department of
Biophysics, University of Rochester,
Rochester, New York 14642 (A.C.N)

INTRODUCTION

The prolactin gene has provided a useful system for
analysis of the mechanism of estrogen action. Estradiol
stimulates prolactin biosynthesis both *in vivo* (Maurer
and Gorski, 1977) and in cultured cells (Haug and
Gautvik, 1976; Lieberman et al., 1978). Estrogen effects
on prolactin biosynthesis are mediated by changes at the
trancriptional level (Maurer, 1982; Shull and Gorski,
1984) which lead to accumulation of prolactin mRNA (Stone
et al., 1977; Ryan et al., 1979; Seo et al., 1979). As
estrogen has been shown to regulate the transcription of
the prolactin gene it is reasonable to suggest that the
estrogen receptor might bind to sequences within or
adjacent to the prolactin gene. The demonstration that
estrogenic stimulation of prolactin gene transcription is
not dependent on protein synthesis (Shull and Gorski,
1984) is consistent with such a direct mechanism for
estrogen effects on prolactin gene expression. Thus
several studies demonstrate that the prolactin gene is
regulated by estrogen and the data are consistent with a
simple model involving direct interaction of the estrogen
receptor with the prolactin gene.

Initial studies have begun to explore the possible
interaction of the estrogen receptor with the prolactin
gene. Studies using crude receptor preparations and a
competition binding assay have demonstrated selective

binding of the estrogen receptor to a region at least one kilobase a upstream from the transcription initiation site rather than to the more proximal 5' flanking region of the gene (Maurer, 1985). Those studies suggested that the region of the prolactin gene responsible for interacting with the estrogen receptor might be rather distinct from the promoter. Interestingly, the upstream region of the prolactin gene has been found to also contain a tissue-specific enhancer element (Nelson *et al.*, 1986) and this region also contains a DNase I hypersensitive site (Durrin *et al.*, 1984).

In the studies described here, highly purified estrogen receptor has been used to examine binding to the 5' flanking region of the prolactin gene. In addition gene transfer studies have been used to test the functional significance of receptor-DNA interaction. The results of these studies have been recently published (Maurer and Notides, 1987; Kim *et al.*, 1988).

ESTROGEN RECEPTOR BINDING TO THE PROLACTIN GENE

The binding of estrogen receptor to fragments of the prolactin gene was examined using a nitrocellulose filter binding assay. This assay takes advantage of the property of nitrocellulose to bind double-stranded DNA only if it is associated with a protein. For the assay, radiolabeled fragments of the prolactin gene were incubated with highly purified estrogen receptor and then filtered through nitrocellulose. DNA which bound to the receptor was then analyzed by agarose gel electrophoresis. The results demonstrated selective binding of receptor to fragments which contain the -1713 to -1530 region of the prolactin gene. The selectivity was similar to that observed for glucocorticoid receptor binding to mammary tumor virus sequences (Payvar *et al.*, 1981; Scheidereit *et al.*, 1983) or progesterone receptor binding to the ovalbumin gene (Compton *et al.*, 1983). The binding demonstrated a strong, non-linear concentration dependence. As receptor concentrations increased from 1.6 to 3.2 nM there was a more than two-fold increase in DNA binding. This non-linear concentration dependence is similar to that observed for procaryotic gene regulatory proteins which are known to act as dimers (Pirrotta *et al.*, 1970; Riggs *et al.*, 1970) and is consistent with the view that dimers may be the physiologically important form of the receptor (Notides et al., 1981).

An effort was then made to further localize the regions of DNA which interact with the receptor. The general region of DNA containing estrogen receptor binding sites was determined through the preparation of 5' and 3' deletions of the prolactin gene. When 5' deletions were prepared, receptor binding activity was suddenly lost when sequences past -1713 were removed. For 3' deletions, binding was reduced when sequences past -1529 were removed. Thus the 5' and 3' deletions suggest that the receptor binding region occurs somewhere between positions -1713 and -1529. To further localize DNA sequences which interact with the receptor, nuclease protection experiments were performed. Incubation of end-labeled DNA representing the -1713 to -1495 region with estrogen receptor was found to protect the DNA from the progressive digestion of exonuclease III. DNA sequences which were protected by the estrogen receptor were found to include positions -1587 to -1563. The sequence in this region of the prolactin gene includes the following sequence and protection region:

```
-1600      -1590       -1580       -1570       -1560         -1550
 .          .      Estrogen Receptor Binding    .             .
TGCATTAAAAAATGCATTTTTGTCACTATGTCCTAGAGTGCTTTGGGGTCA
        ━━━━▶         ◀━━━━
```

The arrows indicate an imperfect palindrome in the rat prolactin gene which is similar to a perfect, conserved palindrome, GGTCANNNTGACC which is found in both the Xenopus and chicken vitellogenin genes. The simple palindromic sequence has been found to be necessary and sufficient for estrogenic regulation of vitellogenin genes (Klein Hitpass et al., 1986; Klock et al., 1987; Martinez et al., 1987) and suggests that the imperfect palindrome might be sufficient to account for estrogenic regulation of the prolactin gene.

GENE TRANSFER STUDIES OF ESTROGEN-RESPONSIVE SEQUENCES

To determine if the sequences which bind the estrogen receptor are actually required to mediate estrogenic regulation of transcription, fusion genes

containing the 5' flanking region of the prolactin gene
linked to the bacterial chloramphenicol acetyltransferase
(CAT) gene were prepared. The fusion genes were
transferred into GH3 pituitary tumor cells and the
response of the CAT marker gene determined. The results
demonstrate that constructs containing more than 1.5
kilobases of 5' flanking DNA show increased expression of
the CAT marker gene in response to estrogen, while
shorter constructs are not estrogen-responsive (Figure
1). Thus the same sequences required for receptor
binding are also required for estrogen-responsiveness.

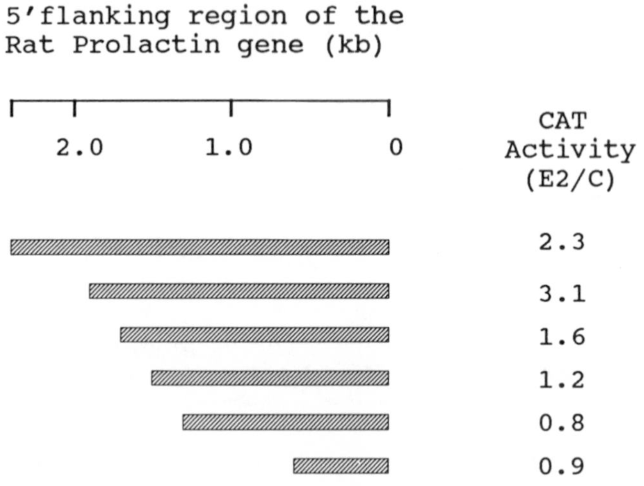

5'flanking region of the
Rat Prolactin gene (kb)

			CAT
2.0	1.0	0	Activity (E2/C)
			2.3
			3.1
			1.6
			1.2
			0.8
			0.9

Figure 1. Analysis of estrogenic regulation of
5'-deleted prolactin-CAT fusion genes. Fusion genes
containing the indicated amount of the 5' flanking region
of the prolactin gene were transfected into GH3 cells in
the absence or presence of 10 nM estradiol. After two
days of exposure to the estradiol, cell extracts were
prepared and assayed for CAT activity. The values shown
are the ratio of activity in estradiol-treated cells to
activity in control cells. Data replotted from Maurer and
Notides (1987).

In the preceding experiment, the fusion gene
constructs contained the prolactin promoter. To
determine if the estrogen-responsive sequences of the
prolactin gene could alter the activity of a heterologous
promoter, a construct containing the promoter of the
herpes simplex virus thymidine kinase (TK) gene was
prepared. For this construct the -1713 to -1495 region
of the prolactin gene was placed in either orientation
upstream of a TK-CAT gene and then transfected into GH3
cells. As expected, the TK-CAT gene by itself was not
estrogen-responsive. Addition of the -1713 to -1495
region of the prolactin gene did confer estrogen-
responsiveness to the TK promoter. Because this small,
approximately 200 base pair fragment was able to permit
estrogen-responsiveness in either orientation, it appears
to function as an estrogen-dependent enhancer sequence.

INTERACTION OF RECEPTOR WITH AN ENHANCER BINDING FACTOR

Because the region of the prolactin gene which
shows selective binding to the estrogen receptor also
contains a tissue-specific enhancer element (Nelson et
al., 1986) and a DNase I hypersensitive site (Durrin et
al. 1984), it seemed likely that factors other than the
estrogen receptor would bind to this region. These
factors might interact functionally with the estrogen
receptor to permit or facilitate the ability of the
receptor to alter transcription. Therefore, we sought
to identify tissue-specific factors which bind to this
region and which might interact with the receptor.

To search for possible interactions with nuclear
factors, a gel mobility shift assay was used (Fried and
Crothers, 1984). After incubation of nuclear extracts
from GH3, JAR, HeLa and CHO cells with a radiolabled DNA
fragment representing the -1713 to -1494 region of the
rat prolactin gene, a slowly migrating DNA-factor complex
was detected only in GH3 cells. The complexes from GH3
cells appeared to be sequence specific as low
concentrations of the unlabeled -1713 to -1494 fragment
treated reduced the amount of the complex. Thus GH3
cells, but not the other cells tested contain a factor
which binds to an upstream fragment of the rat prolactin
gene.

As we have shown that the estrogen receptor binds
to this region of DNA, it was important to determine if
the estrogen receptor might be the factor responsible
for causing the mobility shift. To examine this
possibility, nuclear extracts were prepared from cells
which had been incubated with [^{3}H]estradiol and the
extracts were chromatographed on heparin agarose. Column
fractions were assayed for DNA-binding activity by the
gel mobility shift assay and the estrogen receptor
monitored by determining elution of [^{3}H]estradiol. The
results demonstrated that the activity responsible for
the gel mobility shift is separable from the estrogen
receptor.

To identify at the nucleotide level, the location
of factor binding sites, a nuclease protection experiment
was performed. The results showed that the factor
responsible for the gel mobility shift activity bound to
a rather discrete region of DNA including the sequences:

Protected Area
-1670   CAACTTCATTATTATTCACCATAATGACAT   -1641

The region protected by the GH3 nuclear factor is
indicated. The arrows indicate the location of a
symmetrical sequence. Only one half of the dyad
symmetry appears to be protected by the GH3 nuclear
factor.

The possible significance of the factor-DNA
interaction was evaluated by oligonucleotide-directed
mutagenesis of the factor binding site. As the factor
binding site appeared to involve only half of a dyad
symmetry it seemed appropriate to mutate both the full
dyad as well as the half shown to bind to the factor.
Thus, either the full dyad, or the upstream half were
mutated by T to G transversions of the appropriate
sequences. Gel mobility shift assay demonstrated that
mutation of either the full dyad or the upstream half
eliminated the ability of the factor to bind the DNA. To
test interactions with the estrogen receptor, the region
previously shown to bind the receptor, positions -1582 to
-1569 were also mutated by T to G transversion. Mutation
of the estrogen-response element (ERE) did not alter the

binding of the factor to the prolactin DNA fragment.

The functional significance of disrupting factor-DNA interaction was tested by constructing marker genes containing the -1713 to -1494 region of the prolactin gene coupled to a TK-CAT fusion gene. Either the wild type or mutant fusion gene was transferred to GH3 cells and CAT activity determined (Figure 2). The

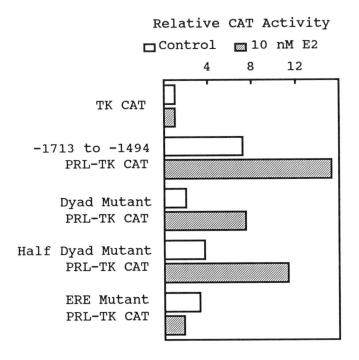

Figure 2. Effects of mutations of prolactin sequences on enhancer activity. Prolactin fragments containing either the wild-type or the indicated mutants were subcloned upstream of the TK-CAT fusion gene. The fusion genes were transferred into GH3 cells by electroporation and the transfected cells received either no treatment or 10 nM estradiol as indicated for 48 hours. Cell extracts were then prepared and assayed for CAT activity. Values are means determined from three transfections. Replotted from Kim et al., 1988.

wild type fragment showed both a substantial increase in basal activity (as compared to TK-CAT) as well as the ability to confer estrogen-responsiveness on the TK promoter. Mutation of either the full dyad, or only the upstream half detectably decreased basal activity, but did not block the ability of estrogen to stimulate activity of the construct. Mutagenesis of the estrogen-receptor binding site had effects on basal activity and totally abolished estrogen effects on marker gene activity.

These findings demonstrate the presence in GH3 nuclear extracts of a factor which appears to bind to a "basal" enhancer element near the estrogen receptor binding site. While binding of the factor is apparently not required for estrogen effects, the mutation studies suggest that the effects of factor binding are additive with estrogen effects. Thus, the data suggest that there are two relatively independent, but functionally interactive elements in this upstream region of the prolactin gene. One element located at positions -1666 to -1659 binds to a tissue-specific factor and is responsible for maintenance of "basal" transcription. The other element located at positions -1582 to -1569 binds the estrogen receptor and is required for estrogen-responsiveness.

Recently Nelson et al. (1988) have identified a factor present in GH3 cell extracts which binds to multiple sites both in the proximal promoter region and in an upstream region. One of the factor binding sites detected by Nelson et al. (1988) appears to be the same site that we have identified in the present study. The findings of Nelson et al. demonstrate that there are multiple factor binding sites in the upstream region near the estrogen receptor binding site. Further studies are required to evaluate the possible role of these factor-DNA interactions in altering the estrogen response of the prolactin gene.

CONCLUSIONS

These studies have identified the DNA sequences which are required for estrogenic regulation of the prolactin gene. Receptor binding and gene transfer approaches have both identified the same DNA sequences. The sequences are similar to estrogen-responsive sequences of vitellogenin genes. It seems likely that at least a portion of the mechanisms involved in estrogenic stimulation of transcription involves interaction of the receptor with sequences related to the palindrome, GGTCANNNTGACC, or minor variations of this sequence. Studies have been initiated to identify factors which might interact with the receptor. To date, a factor which appears to interact with a "basal" enhancer element has been identified. The ability of this enhancer to increase basal transcription appears to be additive with estrogen-stimulated transcription mediated through the estrogen-response element. Further studies may identify other factors which participate in the ability of estrogen to stimulate transcription of prolactin and other genes.

REFERENCES

Compton JG, Schrader WT, O'Malley BW (1983). DNA sequence preference of the progesterone receptor. Proc Natl Acad Sci USA 80: 16-20.

Durrin LK, Weber JL, Gorski J (1984). Chromatin structure, transcription, and methylation of the prolactin gene domain in pituitary tumors of Fischer 344 rats. J Biol Chem 259: 7086-7093.

Fried MG, Crothers DM (1984). Equilibrium studies of the cyclic AMP receptor protein-DNA interaction. J Mol Biol 172: 263-282.

Haug E, Gautvik KM (1976). Effects of sex steroids on prolactin secreting rat pituitary cells in culture. Endocrinology 99: 1482-1489.

Kim KE, Day RN, Maurer RA (1988). Functional analysis of the interaction of a tissue-specific factor with an upstream enhancer element of the rat prolactin gene. Mol Endocrinol, in press.

Klein-Hitpass L, Schorpp M, Wagner U, Ryffel GU (1986). An estrogen-responsive element derived from the 5' flanking region of the Xenopus vitellogenin A2 gene functions in transfected human cells. Cell 46: 1053-1061.

Klock G, Stahle U, Schutz G (1987). Oestrogen and glucocorticoid responsive elements are closely related but distinct. Nature 329: 734-736.

Lieberman ME, Maurer RA, Gorski J (1978). Estrogen control of prolactin synthesis *in vitro*. Proc Natl Acad Sci USA 75: 5946-5949.

Martinez E, Givel F, Wahli W (1987). The estrogen-responsive element as an inducible enhancer: DNA sequence requirements and conversion to a glucocorticoid responsive element. EMBO J 6: 3719-3727.

Maurer RA (1982). Estradiol regulates the transcription of the prolactin gene. J Biol Chem 257: 2133-2136.

Maurer RA (1985). Selective binding of the estradiol receptor to a region at least one kilobase upstream from the rat prolactin gene. DNA 4: 1-9.

Maurer RA, Gorski J (1977). Effects of estradiol-17β and pimozide on prolactin synthesis in male and female rats. Endocrinology 101: 76-84.

Maurer RA, Notides AC (1987). Identification of an estrogen-responsive element from the 5' flanking region of the rat prolactin gene. Mol Cell Biol 7: 4247-4254.

Nelson C., Crenshaw EB III, Franco R, Lira SA, Albert VR, Evans RM, Rosenfeld MG (1986). Discrete cis-active genomic sequences dictate the pituitary cell type-specific expression of rat prolactin and growth hormone genes. Nature 322: 557-562.

Nelson C., Albert VR, Elsholtz HP, Lu LI-W, Rosenfeld MG (1988). Activation of cell-specific expression of rat growth hormone and prolactin genes by a common transcription factor. Science 239: 1400-1405.

Notides AC, Lerner N, Hamilton DE (1981). Positive cooperativity of the estrogen receptor. Proc Natl Acad Sci USA 78: 4926-4930.

Payvar F, Wrange O, Carlstedt-Duke J, Okret S, Gustafsson J-A, Yamamoto KR (1981). Purified glucocorticoid receptors bind selectively in vitro to a cloned DNA fragment whose transcription is regulated by glucocorticoids in vivo. Proc Natl Acad Sci USA 78: 6628-6632.

Pirrotta V, Chadwick P, Ptashne M (1970). Active form of two coliphage repressors. Nature 227: 41-44.

Riggs AD, Suzuki H, Bourgeois S (1970). Lac repressor-operator interactions. J Mol Biol 48: 67-83.

Ryan R, Shupnik MA, Gorski J (1979). Effect of estrogen on preprolactin mRNA messenger ribonucleic acid sequences. Biochemistry 18: 2044-2048.

Scheidereit C, Geisse S, Westphal HM, Beato M (1983) The glucocorticoid receptor binds to defined nucleotide sequences near the promoter of mouse mammary tumor virus. Nature 304: 749-752.

Seo H, Refetoff S, Martino E, Vassart G, Brocas H (1979). The differential stimulatory effect of thyroid hormone on growth hormone synthesis and estrogen on prolactin synthesis due to accumulation of specific messenger ribonucleic acids. Endocrinology 104: 1083-1090.

Shull JD, Gorski J (1984). Estrogen stimulates prolactin gene transcription by a mechanism independent of pituitary protein synthesis. Endocrinology 114: 1550-1557.

Stone RT, Maurer RA, Gorski J (1977). Effect of estradiol-17β on preprolactin messenger ribonucleic acid activity in the rat pituitary gland. Biochemistry 16: 4915-4921.

Molecular Endocrinology and Steroid
Hormone Action, pages 171–186
© 1990 Alan R. Liss, Inc.

TRANSCRIPTION COMPLEXES

Alan P. Wolffe

Laboratory of Molecular Biology, NIDDK,
Bldg. 6, Room 131, NIH, Bethesda, Md. 20892

A transcription complex assembled on the promoter of
a eukaryotic gene has the potential for many functions
aside from directing RNA polymerase to initiate
transcription. The regulation of eukaryotic gene
activity, the stability of the differentiated state, cell
commitment and enhancer function may all be attributed to
properties of these structures.

A eukaryotic gene is transcribed following the
sequential binding of transcription factors to its
promoter, which together assemble a transcription complex
recognized by RNA polymerase. A general problem in
eukaryotic transcription is the significance of the large
number of protein-protein and protein-DNA interactions
required for transcription complex assembly. Prokaryotes
regulate transcription initiation using much simpler
mechanisms (Reznikoff et al., 1985). This paper
considers the structural features of transcription
complexes that might explain some unique aspects of
eukaryotic gene expression.

Assembly of functional transcription complexes

Multiprotein transcription complexes undergo a
highly ordered assembly process at promoters initiated by
sequence-specific DNA binding proteins. Non-DNA binding
proteins are sequestered by virtue of protein-protein
contacts. In some cases, the association of these
proteins alters preexisting DNA-protein interactions both
qualitatively (eg. extent or morphology of DNAase I
footprint) and quantitatively (e.g. the affinity of the
whole complex for the promoter exceeds that of the DNA

binding proteins alone). Over 100 bp of promoter DNA sequence is complexed with multiple proteins on each of the simple promoters illustrated (Figure 1). Interactions between transcription factors over this length of DNA implies either a precise stereospecific orientation of individual proteins on the surface of the DNA helix or that considerable flexibility exists in the structure of the bound proteins. One potential function of the extensive protein-DNA and protein-protein interactions at a eukaryotic promoter is to increase the fidelity of transcription initiation (Echols, 1986). In this model, each specific protein-DNA and protein-protein interaction is a prerequisite for efficient transcription, and provides a reference point for RNA polymerase in aligning the initiation site of transcription.

The complex multisubunit RNA polymerases have not been shown to have a role in facilitating the assembly of transcription complexes. The fact that they have high affinity for proteins associated with transcription complexes implies that certain transcription factors may associate with RNA polymerase in the absence of DNA. This has been shown to be the case for the general class III gene transcription factors TFIIIB and TFIIIC (Wingender et al., 1986), and for RNA polymerase II transcription factors. Some of these proteins may always be associated with RNA polymerase in vivo, so the boundary between being a component of the transcription complex or of polymerase becomes artificial. After transcription initiation and promoter clearance by RNA polymerase, some of the non-DNA binding transcription factors may dissociate, preventing reinitiation by the enzyme until they have rebound (Van Dyke et al., 1988; Hai et al., 1988). The DNA binding proteins remain, these may be responsible for the phenomenon of template commitment, the first function of a transcription complex unique to eukaryotes that we will consider.

Stable commitment to gene activity

The stable sequestration of transcription factors has been described for representatives of all classes of eukaryotic gene (Brown, 1984). The assay for stable transcription complex formation involves the sequential addition of two genes to an in vitro transcription extract. If the first gene added binds all the limiting transcription factors, then the second gene will not be transcribed. If this is true in vivo, then stable

Figure 1. Transcription complexes on eukaryotic genes.

The regions of promoters associated with transcription factors are indicated for all three classes of eukaryotic gene. The solid bar indicates the gene itself, numbers indicate the boundaries of the DNase I footprints of either purified DNA binding proteins or of the complete transcription complex.

A.   The Xenopus somatic 5S RNA gene transcription complex (class III) (Wolffe and Brown, 1988).

B.   The adenovirus major late promoter transcription complex (class II) (Van Dyke et al., 1988).

C.   The human ribosomal RNA gene transcription complex (class I) (Bell et al., 1988).

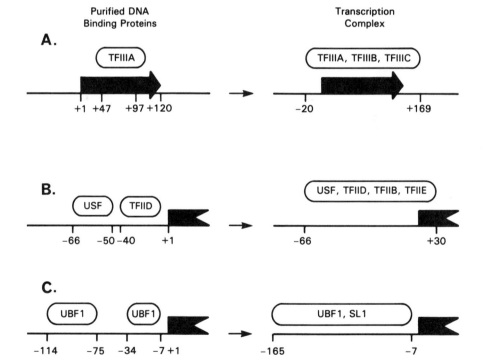

transcription complexes might explain the maintenance of distinct patterns of gene activity in a terminally differentiated cell (Bogenhagen et al., 1982). Recent experiments support this idea, in a living Xenopus oocyte nucleus, transcription complexes assembled onto a somatic 5S RNA gene are stable. Furthermore, somatic 5S RNA gene transcription complexes can also be found in chromatin isolated from cells in which there are no longer free transcription factors (Darby et al., 1988). The stability of a transcription complex depends on the multiple protein-protein and protein-DNA interactions involved in its assembly. This stability is distinct from the transition seen in prokaryotic systems from the unstable closed complex to the stable, transcriptionally active, open complex in which DNA is unwound at the promoter (Hawley and McClure 1982; Hayes et al., 1989). The requirement for many interactions to generate a stable complex affords multiple opportunities for regulating these interactions and thereby modulating gene activity (see below).

An active eukaryotic gene has to maintain a transcription complex through a variety of potentially disruptive events. Transcription and replication of the gene will be discussed later, however the transcription complex remains in place in spite of chromatin assembly and the compaction of chromosomes at mitosis. The physical state of chromatin surrounding active and repressed genes differs in respect to its accessibility to enzymes such as DNase I (Weintraub, 1985). Active chromatin is more accessible, this may reflect the inhibition of chromatin compaction caused when stable transcription complexes interrupt regular nucleosome formation. A transcription complex provides a highly 'visible' signal for RNA polymerase in a background of 'invisible' DNA packaged into nucleosomes. In fact without transcription complexes, it might be argued that all the DNA in eukaryotic promoters would be 'invisible' as compacted chromatin. If a transcription complex is not stable (see later) changes in chromatin structure may direct the dissociation of transcription factors and the repression of genes (Wolffe, 1989). Chromatin assembly should therefore not be thought of as having a merely passive role in the sequestration of DNA in the eukaryotic nucleus.

Regulation of gene activity.

Stable transcription complexes allow a gene to be active indefinitely, however many gene systems are regulated. For example, a gene that needs to be inactivated during development might make use of a transcription complex that is unstable. The gene would be active at high concentrations of transcription factors but inactive when levels fell below a certain threshold. This type of regulation is seen with the 5S RNA genes of Xenopus laevis, where the somatic 5S RNA genes form stable transcription complexes, whereas the oocyte 5S RNA genes do not (Wolffe and Brown, 1987; Wolffe, 1988). During embryogenesis the oocyte 5S RNA genes are turned off, whereas the somatic 5S RNA genes remain active (Wormington and Brown, 1983; Wakefield and Gurdon, 1983). A reduction in transcription factor concentration during embryogenesis leads to the selective dissociation of oocyte 5S RNA gene transcription complexes. Furthermore chromatin assembly both prevents transcription factors reassociating with the oocyte 5S DNA and may direct the dissociation of transcription factors from genes leading to repression (Schlissel and Brown, 1984; Wolffe, 1989). The DNA sequence differences responsible for this differential regulation of the 5S RNA genes appear to consist of only three base pairs. However two transcription factors bind to this region of the gene (Pieler et al., 1987; Wolffe, 1988). Changes in the binding affinity of each of the two proteins amplify their individual effect on complex stability. A difference in gene activity of over 1000 fold can therefore be explained simply by differences in the stability of protein-protein and protein-DNA interactions in oocyte or somatic 5S RNA gene transcription complexes.

It may be generally true that regulated genes will rely on weak protein-protein or protein-DNA interactions to retain flexibility in transcriptional activity. Individual eukaryotic transcription factors often bind weakly to their cognate DNA sequences; the ease with which these factors can be dissociated contributes to gene regulation. The binding of steroid receptors to target genes is a good example. The glucocorticoid receptor binds to the mouse mammary tumor virus promoter displacing a nucleosome, this facilitates the assembly of a transcription complex (Richard Foy and Hager, 1987). When the concentration of hormone falls, the glucocorticoid receptor dissociates from the promoter and transcription stops. Genes that are always active in a particular tissue may have to rely on much higher affinity protein-DNA interactions to maintain stable transcription complexes.

Transcription complex stability during transcription and replication

Two related functions of transcription complexes have now been considered: genes that need to be continually active require stable complexes, while genes that need to be regulated utilize unstable complexes. However, once a transcription complex is assembled its properties may be influenced by the passage of RNA and DNA polymerases. The consequences of these potentially disruptive processive enzymes passing through a transcription complex are discussed below.

An immediate problem in transcribing a class III gene is that all of the essential promoter elements are within the gene sequence (Ciliberto et al., 1983). On a somatic 5S RNA gene, transcription factors associated with these sequences remain stably bound in spite of hundreds of transits by RNA polymerase III (Wolffe et al., 1986). Experiments with bacteriophage RNA polymerase revealed that the presence of multiple DNA binding proteins coupled together by protein-protein contacts allows individual proteins to anchor the complex to DNA. Transient dissociation of any one contact need not lead to dissociation of the whole complex.

A very different result is seen when only a single transcription factor is associated with a promoter, as in the case of the Acanthamoeba class I ribosomal RNA gene. RNA polymerase passage through this promoter leads to the displacement of the single transcription factor (Bateman and Paule, 1988). This may explain the significance of transcription termination sites being placed upstream of the promoters of genes arranged in tandem arrays (e.g rDNA, McStay and Reeder, 1986). Promoter occlusion by a transcribing RNA polymerase is well known on prokaryotic genes (Adhya and Gottesman, 1982; Horowitz and Platt, 1982).

The capacity to maintain protein-DNA interactions in place during transcription may be an important element of transcription complex structure contributing to the regulation of many genes. Some class II genes are also known to have regulatory elements and protein-DNA complexes within either exons and introns (Banerji et al., 1983; Theulaz et al., 1988). The capacity to have stable complexes assembled downstream of the promoter contributes another dimension of flexibility to eukaryotic gene regulation (Schaffner et al., 1988). The adenovirus major late promoter directs RNA polymerase to transcribe a gene which contains five other active promoters including stable complexes assembled on both the class III VA genes and class II promoters (Berk, 1986). The maintenance of a transcription complex in spite of transcription through it (Wolffe et al., 1986), means that we should not perhaps be too surprised to find overlapping transcription units in the eukaryotic genome.

A related problem of maintaining specific protein-DNA interactions associated with a transcription complex occurs when a replication fork passes along a gene. What happens to the transcription factors comprising the complex may have important implications for the inheritance of patterns of gene activity in eukaryotic cells.

Experiments that attempt to test the maintenance of transcription complexes through replication in vivo have generally made use of 'enhancer dependent' promoters. As will be discussed later, enhancers appear to facilitate the assembly of transcription complexes on certain promoters. In one experiment, Calame and colleagues established competition for simian virus 40 (SV40) enhancer factors, after enhancer dependent transcription had been initiated on a gene (Wang and Calame, 1985). Transcription from the promoter of the gene continued in spite of the competition. Moreover, replication of the transcriptionally active gene did not inhibit transcription even in the presence of the competitor DNA. This suggests that the transcription complex on the promoter was stable to replication fork passage.

This particular problem has often been investigated and discussed using immunoglobulin genes as examples. These genes alter their regulatory elements during lymphoid cell differentiation. The immunoglobulin heavy chain (IgH) enhancer is required to activate transcription from IgH promoters early in B-cell differentiation (Banerji et al., 1983). However, several differentiated B-lymphoid cell lines exist that have deleted the IgH enhancer, but retain normal levels of IgH transcription (Wabl and Burrows, 1984; Klein et al., 1985). There are several possible explanations for this result (Figure 2). One of these is that the IgH enhancer is required only for the establishment of the IgH promoter transcription complex early in B-cell differentiation, later on the enhancer can be deleted and the transcription complex will remain in place in spite of cell division. Alternatively, the enhancer is required for maintenance of the IgH gene transcription complex during cell division, but when deleted can be replaced by other regulatory elements (Grosschedl and Marx, 1988). A third explanation might be that some other modification of active chromatin such as demethylation, which can be propagated at the replication fork, might explain continued gene activity (Kelley et al., 1988). Definitive proof of the stability of a transcription complex through cell division requires the establishment of an in vitro system, where the physical structure of transcription complexes assembled on a particular promoter can be analyzed before and after DNA replication.

Evidence that argues against the general maintenance of transcription complexes on all genes during replication comes from in vitro experiments in which the physical structure of a 5S RNA gene transcription complex was analyzed before and after replication (Wolffe and Brown, 1986). In contrast to the stability of the transcription complex to transcription, replication fork progression disrupted the complex and displaced transcription factors. No selective advantage existed for rebinding factors to the daughter 5S RNA genes, that had initially had a transcription complex, compared to naked 5S DNA. Constitutively expressed genes such as the 5S RNA gene may not require stability to replication, this property may be restricted to the complexes of tissue specific genes.

A role for the extensive cooperative, protein-protein and protein-DNA interactions in maintaining transcription complex structure following replication therefore remains to be proved. It is particularly exciting that these structures might be maintained, thereby providing a molecular explanation for the phenomenon of cell commitment during embryonic development (Brown, 1984; Weintraub, 1985).

Enhancers and the transcription complex

Many eukaryotic genes require enhancers for maximal gene activity (Serfling et al., 1985). Although the precise mechanism of enhancer action is not understood (Ptashne, 1986), enhancer mediated transcription seems to involve facilitating the establishment of transcription complexes (Mattaj et al., 1985; Weintraub, 1988). There are many similarities in the assembly of the nucleoprotein structures at both eukaryotic promoters and enhancers. Both complexes are made up of multiple sequence elements, each of which binds a cognate transcription factor (Zenke et al., 1986; Wildeman et al., 1986; Fromental et al., 1988). Both may require stability to either transcription or replication (Schaffner et al., 1988; Wang and Calame, 1985). Stable enhancer complexes may be important in maintaining tissue specificity, even though the activity of particular genes may change (Choi and Engel, 1988). Unstable enhancer complexes are important in regulating gene activity. For example, the mouse mammary tumor virus enhancer is only active when glucocorticoid receptor is bound (Yamamoto, 1985). Removal of the steroid hormone results in transcriptional inactivation, indicating that the glucocorticoid receptor is required for both establishment and maintenance of enhancer mediated effects.

Many DNA binding proteins are shared between enhancer sequences and promoters (Falkner and Zachau, 1984; Bienz and Pelham, 1986; Evans et al., 1988), conceivably non-DNA binding proteins will be also be shared. The interaction of a non-DNA binding protein with a DNA binding protein at two sites provides a simple explanation for the possible looping between enhancer and promoter elements. (Figure 2ii, Ptashne, 1986). The distinction between transcription complexes and enhancer complexes may in fact be artificial. A single structure combining both elements, affords much greater possibilities for each of the potential functions discussed above. For instance, one reason for the separation of enhancers and promoters on DNA over extensive distances may be that any one structure might be disrupted by DNA replication, while the other would remain intact (Figure 2iv). If protein binding to one sequence element influences the binding of proteins to the other, then the intact nucleoprotein complex might facilitate the reformation of the disrupted one.

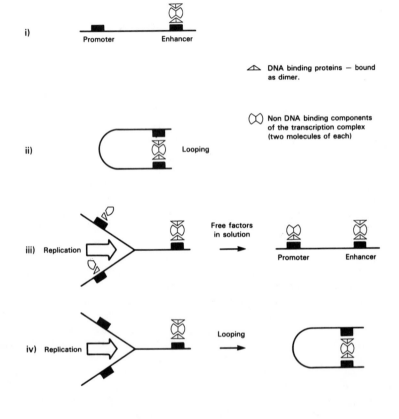

i)

Promoter   Enhancer

⚬ DNA binding proteins — bound
   as dimer.

⚬ Non DNA binding components
   of the transcription complex
   (two molecules of each)

ii)   Looping

iii) Replication   Free factors
                    in solution

Promoter   Enhancer

iv) Replication   Looping

Figure 2. Models for the maintenance of transcription complexes through replication.

The regulatory regions of a gene are shown as a promoter and an enhancer (bars). Open boxes are DNA binding or non DNA binding components of a transcription complex. (i) In this case similar factors are shared between enhancers and promoters. (ii) The sequestration of transcription factors onto the promoter (and its activation) is facilitated by looping of the intervening DNA between enhancer and promoter. Common DNA binding proteins associate with enhancer and promoter. The bound proteins interact with a common non-DNA binding protein. (iii) DNA replication disrupts the transcription complex on the promoter splitting it in half. The remaining transcription factors can sequester free factors from solution generating a complete transcription complex on each daughter chromatid. In this case the enhancer is required for establishment, but not maintenance of the transcription complex. (iv) Alternatively DNA replication again disrupts the complex on the promoter, displacing transcription factors, however because of the distance between enhancer and promoter, the enhancer complex remains intact. DNA looping can establish a new transcription complex on the promoter. Here, the enhancer is required for both establishment and maintenance of the transcription complex through cell division.

Conclusions

Eukaryotic transcription complexes have the essential role of directing the accurate and efficient initiation of transcription by RNA polymerase on a particular gene. The focus of much current research in molecular biology lies in understanding the regulation of the frequency with which RNA polymerase initiates transcription. The multiplicity of proteins and protein-DNA interactions assembling transcription complexes of different stabilities afford many opportunities for regulating complex structure and therefore transcription initiation itself. For example, this could occur through combinatorial effects of multiple proteins binding to particular DNA sequences or by transiently associating and dissociating a particular transcription factor.

Transcription complex structure may contribute to several other features characteristic of eukaryotic gene expression. The stable sequestration of transcription factors onto a gene by virtue of cooperative interactions between individual factors, can explain the terminal differentiation of a cell type and the stability of a pattern of gene activity over long periods of time. This may be helped by the stability of a complex to the process of transcription itself. The commitment of a gene to a continued state of activity in a particular cell lineage might also be explained by cooperative interactions between components of a transcription complex and maintenance of the structure through DNA replication and cell division. Aside from simply directing RNA polymerase to the gene, many of the exciting and unresolved problems associated with determination and differentiation may be approached by considering the properties of multi-protein transcription complexes on eukaryotic genes.

ACKNOWLEDGEMENTS

I thank Drs. Randall Morse, Tom Sargent and Elizabeth Wolffe for their helpful comments on the manuscript and Thuy Ngo for its preparation.

REFERENCES

Adhya, S., and Gottesman, M. (1982). Promoter occlusion: transcription through a promoter may inhibit its activity. Cell 29 939-944

Banerji, J., Olson, L. and Schaffner, W. (1983). A lymphocyte - specific cellular enhancer is located downstream of the joining region in immunoglobulin heavy chain genes. Cell 33 729-740

Bateman, E., and Paule, M.R. (1988). Promoter occlusion during ribosomal RNA transcription. Cell 54 985-992

Bell, S.P., Learned, R.M., Jantzen, H.M. and Tjian, R. (1988). Functional cooperativity between transcription factors UBF1 and SL1 mediates human ribosomal RNA synthesis. Science 241 1192-1197

Berk, A.J. (1986). Adenovirus promoters and E1A transactivation. Ann. Rev. Gent. 20 45-79

Bienz, M. and Pelham, H.R.B. (1986). Heat shock regulatory elements function as an inclucible enhancer in the Xenopus hsp 70 gene and when linked to a heterologous promoter. Cell 45 753-760

Bogenhagen, D.F., Wormington, W.M., and Brown, D.D. (1982). Stable transcription complexes of Xenopus 5S RNA genes : a means to maintain the differentiated state. Cell 28 413-421

Brown, D.D. (1984). The role of stable complexes that repress and activate eucaryotic genes. Cell 37 359-365

Choi, O-R.B. and Engel, J.D. (1988). Developmental regulation of ϵ globin gene switching. Cell 55 17-26

Ciliberto, G., Castagnoli, L. and Cortese, R. (1983). Transcription by RNA polymerase III. In current topics in developmental biology, Vol. 18 (New York: Academic Press), pp. 59-88

Darby, M.K., Andrews, M.T. and Brown, D.D. (1988). Transcription complexes that program Xenopus 5S RNA genes are stable in vivo. Proc. Natl. Acad. Sci. USA 85 (in press).

Echols, H. (1986). Multiple DNA-protein interactions governing high precision DNA transactions. Science 233 1050-1056

Evans, T., Reitman, M., and Felsenfeld, G. (1988). An erythrocyte specific DNA binding factor recognizes a regulatory sequence common to all chicken globin genes. Proc. Natl. Acad. Sci. USA 85 5976-5980

Falkner, F.G. and Zachau, H.G. (1984). Correct transcription of an immunoglobulin K gene requires an upstream fragurent containing conserved sequence elements. Nature 310 71-74

Fromental, C., Konoo, M., Nomiyama, H., and Chambon, P. (1988). Cooperativity and hierarchical levels of functional organization in the SV40 enhancer. Cell 54 943-953

Grosschedl, R. and Marx, M. (1988). Stable propagation of the active transcriptional state of an immunoglobulin μ gene requires continuous enhancer function. Cell 55 645-654

Hai T., Horikoshi, M., Roeder, R.G. and Green, M.R. (1988). Analysis of the role of the transcription factor ATF in the assembly of a functional preinitiation complex. Cell 54 1043-1051

Hawley, D.K. and McClure, W.R. (1982). Mechanism of activation of transcription initiation from the λPrm promoter. J. Mol. Biol. 157 493-525

Hayes, J., Tullius, T.D., and Wolffe, A.P. (1989). A protein-protin interaction is essential for stable complex formation on a 5S RNA gene. J. Biol. Chem. 264 6009-6012

Horowitz, H., and Platt, T. (1982). Regulation of transcription from tandem and convergent promoters. Nucl. Acids. Res. 10 5447-5465

Kelley, D.E., Pollock, B.A., Atchison, M.L. and Perry, R.T. (1988). The coupling between enhancer activity and hypomethylation of k immunoglobulin genes is developmentally regulated. Mol. Cell Biol. 8 930-937

Klein, S., Gerster, T., Picard, D., Radbruch, A. and Schaffner, W. (1985). Evidence for transient requirement of the IgH enhancer. Nucl. Acids Res. 13 8901-8912

Mattaj, I., Lienhard, S., Jiricny, J., and De Roberis, E. (1985). An enhancer - like sequence within the Xenopus U2 gene promoter facilitates the formation of stable transcription complexes. Nature 316 163-167

McStay, B., and Reeder, R.H. (1986). A termination site for Xenopus RNA polymease I also acts as an element of an adjacent promoter. Cell 47 913-920

Pieler, T., Hamm, J. and Roeder, R.G. (1987). The 5S internal control region is composed of three distinct sequence elements, organized as two functional domains with variable spacing. Cell 48 91-100

Ptashne, M. (1986). Gene regulation by proteins acting nearby and at a distance. Nature 322 697-701

Reznikoff, W.S., Siegele, D.A., Cowing, D.W. and Gross, C.A. (1985). The regulation of transcription initiation in bacteria. Ann. Rev. Geret. 19 355-387

Richard-Foy, H., and Hager, G.L., (1987). Sequence specific positioning of nucleosomes over the steroid-inducible MMTV promoter. EMBO J. 6 2321-2328

Schaffner, G., Schirm, S., Muller-Baden, B., Weber, F., and Schaffner, W. (1988). Redundancy of information is enhancers as a principle of mammalian transcriptional control. J. Mol. Biol. 201 81-90

Schlissel, M.S. and Brown, D.D. (1984). The transcriptional regulation of Xenopus 5S RNA genes in chromatin : the roles of active stable transcription complexes and histone H1. Cell 37 903-911

Segall, J., Matsui, T. and Roeder, R.G. (1980). Multiple factors are required for the accurate transcription of purified genes by RNA polymerase III. J. Biol. Chem. 255 11986-11991

Serfling, E., Jasin, M. and Schaffner, W. (1985). Enhancers and eukaryotic gene transcription. Trends Genet 1 224-230

Theulaz, I., Hipskind, R., TenHeggeler-Bordier., B., Green, S., Kumar, V., Chambon, P., and Wahli, W., (1988). Expression of human estrogen receptor mutants in Xenopus oocytes : correlation between transcriptional activity and ability to form protein-DNA complexes. The EMBO J. 7 1653-1660

Van Dyke, M., W., Roeder, R.G. and Sawadogo, M. (1988). Physical analysis of transcription preinitiation complex assembly on a class II gene promoter. Science 241 1335-1338

Wabl, M.R., and Burrows, P.D. (1984). Expression of immunoglobulin heavy chain at a high level in the absence of a proposed immunoglobulin enhancer in cis. Proc. Natl. Acad. Sci. USA. 81 2452-2455

Wakefield, L., and Gurdon, J.B. (1983). Cytoplamic regulation of 5S RNA genes is nuclear-transplant embryos. EMBO J. 2 1613-1619

Wang, X.F., and Calame, K. (1986). SV40 enhancer-binding factors are required at the establishment but not the maintenance step of enhancer-dependent transcriptional activation. Cell 47 241-247

Weintraub, H. (1985). Assembly and proagation of repressed and derepressed chromosomal states. Cell 42 705-711

Weintraub, H. (1988). Formation of stable transcription complexes as assayed by analysis of individual templates. Proc. Natl. Acad. Sci. USA 85 5819-5823

Wildeman, A.G., Zenke, M., Schatz, C., Wintzerith, M., Grundstrom, T., T. Matthes, H., Takahaski, K. and Chambon, P. (1986). Specific protein binding to the simian virus 40 enhancer in vitro. Mol. Cell. Biol. 6 2098-2105

Wingender, E., Jahn D. and Seifart, K.H. (1986). Association of RNA polymerase III with transcription factors in the absence of DNA. J. Biol. Chem. 261 1409-1413

Wolffe, A.P. (1988). Transcription fraction TFIIIC can regulate differential Xenopus 5S RNA gene transcription in vitro. The EMBO J. 7 1071-1079

Wolffe, A.P. (1989). Dominant and specific repression of Xenopus oocyte 5S RNA genes and satellite I DNA by histone H1. The EMBO J. 8 527-537

Wolffe, A.P. and Brown, D.D. (1986). DNA replication in vitro erases a Xenopus 5S RNA gene transcription complex. Cell 47 217-227

Wolffe, A.P., and Brown, D.D. (1987). Differential 5S RNA gene expression in vitro. Cell 51 733-740

Wolffe, A.P., and Brown, D.D. (1988). Developmental regulation of two 5S ribosomal RNA genes. Science 241 1626-1632

Wolffe, A.P., Jordan, E. and Brown, D.D. (1986). A bacteriophage RNA polymerase transcribes through a Xenopus 5S RNA gene transcription complex without disrupting it. Cell 44 381-389

Wormington, W.M., and Brown, D.D. (1983). Onset of 5S RNA gene regulation during Xenopus embryogenesis. Dev. Biol. 99 248-257

Yamamoto, K.R. (1985). Steroid receptor regulated transcription of specific genes and gene networks. Ann. Rev. Genet. 12 209-252

Zenke, M., Grundstrom, T., Matthes, H., Wintzerith, M., Schatz, C., Wildeman, A. and Chambon, P. (1986). Multiple sequence motifs are involved in SV40 enhancer function. The EMBO J. 5 387-397

Molecular Endocrinology and Steroid
Hormone Action, pages 187–197
© 1990 Alan R. Liss, Inc.

ESTROGEN REGULATION OF TRANSCRIPTION

Deborah A. Lannigan and Angelo C. Notides

EHS Center, University of Rochester Medical
School, Rochester, NY 14642

INTRODUCTION

Within a responsive cell, estrogen binds reversibly but with high affinity to a specific receptor (Gorski et al., 1968; Jensen and DeSombre, 1973; Skafar and Notides, 1987). This estrogen-receptor complex then alters the transcription rate of a battery of estrogen responsive genes (Katzenellenbogen and Gorski, 1975). Molecular models of the mechanism for estrogen regulation of transcription have implicated the involvement of RNA (Kumar and Dickerman, 1985), nonhistone proteins (Toft et al., 1987; Feavers et al., 1987; Redevilh et al., 1987) and receptor phosphorylation (Auricchio et al., 1987) with only slight emphasis on the role the steroid plays. Obviously, however, the steroid should be the primary vector of information to alter gene transcription. This review presents the argument that estrogen alone both activates its receptor, and controls the interaction with DNA which results in selective gene activation. Additionally, we argue that the allosteric properties of the estrogen receptor provide a sensitive biological control mechanism.

PURIFICATION AND QUATERNARY STRUCTURE OF THE ESTROGEN RECEPTOR

Crucial to our studies was the purification of the estrogen receptor. Estrogen receptor can be purified from calf uteri to near homogeneity by ammonium sulfate precipitation, estradiol affinity and heparin chromatography (Notides et al., 1985). The purified receptor appears as a single band when 2 μg is electrophoresized on a SDS polyacrylamide gel and stained with Coomassie Blue (0.05 μg of protein can be detected by this method). The specific activity of the purified receptor is equivalent to one mole of estradiol bound per 67,800 gm of receptor or to the gram molecular weight of each estradiol binding subunit.

Three forms of the receptor are known to exist (Notides and Nielsen, 1974; Notides and Nielsen, 1975; Notides and Sasson, 1983): a monomer, a homodimer and a homodimer form associated with nonhormonal binding components. The monomeric form is found only weakly associated with nuclear sites (readily extracted by buffer without KCl) whereas the dimeric form is tightly associated with nuclear sites (extracted only with high salt buffer). The dimeric form associated with additional components is only observed in crude cytosol following sucrose gradient analysis in the absence of salt. The biological significance, if any, of this form remains to be defined.

COOPERATIVE STEROID BINDING

Equilibrium binding analysis of estradiol with the estrogen receptor demonstrates that estradiol binding to the receptor is a positively cooperative interaction (Notides et al., 1981). This cooperativity is demonstrated by the convex Scatchard plot of [^{3}H]estradiol binding to the receptor and a Hill coefficient (an index of cooperativity) of approximately 1.6. Both the evidence for cooperativity and the molecular weight analysis (which indicates that there is only one estradiol-binding site per monomer) strongly support the concept that above a 1 nM concentration the estrogen receptor exists as a homodimer. However, below a receptor concentration of 1 nM the cooperativity decreases which suggests that the dimer dissociates into the monomeric form of the receptor (Notides et al., 1981). The intracellular concentration of receptor is approximately 10 to 40 nM (Notides et al., 1985). This concentration would be much higher if the receptor were exclusively nuclear. Therefore, the physiological form of the receptor is predominately the homodimer and cooperative steroid binding will exist in vivo.

ESTRADIOL-INDUCED CONFORMATIONAL CHANGES OF THE RECEPTOR

The effect of dimerization and steroid binding on the affinity of the receptor for DNA was measured using DNA-sepharose column chromatography (Skafar and Notides, 1988). The receptor partitions between the DNA and the elution buffer. The higher the affinity of the receptor for DNA, the less receptor is eluted from the column in each fraction. Therefore, the slope of the elution profile is a measure of the affinity of the receptor for DNA. The steriod-bound dimer has a 5 to 10-fold greater affinity for DNA than the unliganded dimer. Additionally, the unliganded dimer and the steroid-bound and unliganded monomer have approximately the same DNA affinity. Thus neither monomer: monomer interaction nor estradiol binding to the monomer alter DNA affinity. Only when estradiol binds to the dimer is an increase in affinity of the receptor for DNA observed. These results provide evidence that the increase in affinity of the steroid-bound dimer for DNA is due to an estradiol-induced conformational change of the dimer form.

Further characterization of the changes in the DNA-binding site induced by estradiol-binding was obtained by examining the effect of pH and salt concentration on the affinity of the steroid-bound and unliganded dimeric receptor for DNA (Skafar and Notides, 1988). The affinity of the steroid-bound dimeric receptor for DNA is maximum at pH 7.4 and decreases with increasing pH, until at pH 8.0 the affinities for DNA of both steroid-bound and unliganded dimeric receptor are equal. In contrast, the affinity of the unliganded dimeric receptor for DNA is independent of pH. The number of salt bridges between the receptor and DNA can be determined from the dependence of the affinity on salt concentration (Skafar and Notides, 1988). The estradiol-bound dimeric receptor makes 12-14 salt bridges with DNA, whereas the unliganded dimeric receptor makes only 8 salt bridges. These differences provide further evidence that a hormone-dependent conformational change occurs in the receptor which is responsible for the increased affinity of the steroid-bound dimeric receptor for DNA compared with that of the unliganded dimeric receptor.

SEQUENCE SELECTIVITY AND COOPERATIVE BINDING OF RECEPTOR WITH DNA

Several laboratories have defined a specific DNA sequence (termed estrogen responsive element, ERE) as the basis of selectivity that limits estrogen action to particular genes (Klein-Hitpass et al., 1986; Martinez et al., 1987; Maurer and Notides, 1987; Klock et al., 1987; Waterman et al., 1988; Burch et al., 1988; Klein-Hitpass et al., 1988; Metzger et al., 1988). The ERE consensus sequence, GGTCANNNTGACCT, was derived from the rat prolactin gene and the vitellogenin gene from Xenopus and from chicken by gene transfer experiments and receptor-DNA binding studies. The palindromic ERE sequence behaves like an enhancer in that it operates on a cis-linked promoter at great distances in an orientation-independent manner (Martinez et al., 1987).

The binding of estrogen receptor to ERE-containing fragments from the rat prolactin gene was examined using a nitrocellulose filter binding assay (Maurer and Notides, 1987). This assay is based on the fact that nitrocellulose binds double-stranded DNA only if it is associated with a protein. The estrogen receptor bound to the ERE-containing fragments in a non-linear, concentration-dependent manner. This result demonstrates that the estrogen receptor is binding as a dimer to its target site in a positively cooperative manner.

BINDING MECHANISM

Estrogen responsiveness can be conferred by the perfect palindromic ERE sequence alone (Klock et al., 1987; Martinez et al., 1987; Klein-Hitpass et al., 1988). Point mutations in this sequence drastically alter the steroid responsiveness and selectivity (Klock et al., 1987; Martinez et al., 1987; Klein-Hitpass et al., 1988). However, in vivo, imperfect palindromic EREs are known to exist and appear to confer estrogen responsiveness (Maurer and Notides, 1987; Martinez et al., 1987; Shupnik et al., 1988; Burch et al., 1988; Waterman et al., 1988). Moreover, the homology of the various EREs to the consensus ERE sequence is only slightly greater than to the consensus sequence of the glucocorticoid responsive element, yet no overlap in steroid responsiveness exists (Klock et al., 1987; Martinez et al., 1987; Klein-Hitpass et al., 1988). Therefore, we investigated whether parameters other than the sequence of the ERE could aid in determining the selectivity of the estrogen receptor for DNA. Using the gel mobility assay (Fried and Crothers, 1981; Hendrickson, 1985) to detect specific protein-DNA interactions we analyzed purified estrogen receptor binding to the ERE contained on a 255 bp restriction fragment which was isolated from the upstream region between nucleotides -1784 to -1530 of the rat prolactin gene (Lannigan and Notides, 1988).

Receptor binding to the double-stranded 255 bp fragment was not detected at intracellular concentrations of estrogen receptor under low stringency conditions. However, we also tested receptor binding to the 255 bp fragment after it had been heated to dissociate the double-stranded DNA and then cooled rapidly in order to trap unusual secondary structure. Incubation of the heat-treated fragment with receptor resulted in the appearance of many retarded bands. Under high stringency conditions only two of these retarded bands remained which indicated that these bands represent receptor-DNA interactions of high affinity.

Heat treatment causes the strands of the 255 bp fragment to separate. These single-strands have a decreased gel mobility compared to the double-stranded 255 bp fragment (in the absence of estrogen receptor). The decrease in mobility is anomolous since a similar sized random piece of DNA increased its mobility upon heat treatment. This anomolous mobility suggests that the separated strands of the 255 bp fragment contain secondary structure.

Using single-stranded [^{32}P] end-labeled 255 bp fragment, estrogen receptor binding was detected only to the "coding strand" (strand which is the template for RNA) and not to the "noncoding strand". This result serves as a control demonstrating that the receptor does not bind to all single-strand DNA. Since the prolactin ERE is almost palindromic, the specificity of the receptor for the "coding strand" versus the "noncoding strand" is remarkable.

As mentioned previously, two retarded bands appeared upon incubation of the radiolabelled coding strand of the

255 bp fragment with estrogen receptor. This result suggests the existence of two distinct sites on the 255 bp fragment, such that the receptor could bind to either or both sites. Inspection of the nucleotide sequence (Maurer, 1985) of the 255 bp fragment reveals, in addition to the known ERE, a nucleotide sequence (located between nucleotides -1722 to -1709) with an 80 percent homology to the consensus ERE. Additionally, a third retarded band was revealed on longer exposures of the autoradiograms which would correspond to receptor bound to both sites. The two binding sites, the known ERE and the putative second ERE, were confirmed by competition experiments using subfragments of the 255 bp fragment. The receptor has approximately a 4-fold higher affinity for the putative second ERE than the known ERE. The affinity of the estrogen receptor for the putative second ERE is approximately 10^{11} M^{-1}.

The hypothesis that the putative ERE appears to be functional in vivo is based on the observations that the putative ERE falls in one of two regions crucial for full enhancer function for the prolactin gene in lactotrophs (Nelson et al., 1986), and that a decrease in estrogen responsiveness of a prolactin CAT (chloramphenicol acetyltransferase) fusion gene is observed when the region containing the putative ERE is deleted (Maurer and Notides, 1987).

To determine the specificity of the receptor for the "coding strand" ERE the relative affinities of receptor for various DNAs were measured by competition experiments. The estrogen receptor has a 20-fold lower affinity for single-stranded nonspecific (208 bp fragment without EREs isolated from pUC13) DNA; a 60-fold lower affinity for double-stranded specific (255 bp fragment with EREs) DNA and a 1000-fold less affinity for double-stranded nonspecific DNA compared with its affinity for single-stranded specific DNA. Thus the affinity of receptor for double-stranded specific DNA is 17-fold greater than for double-stranded nonspecific DNA.

MODEL FOR IN VIVO ENHANCEMENT OF TRANSCRIPTION BY ESTROGEN

We propose that in vivo (Figure 1) the receptor exists as a homodimer of two identical monomers, each monomer containing a single estrogen binding site and a

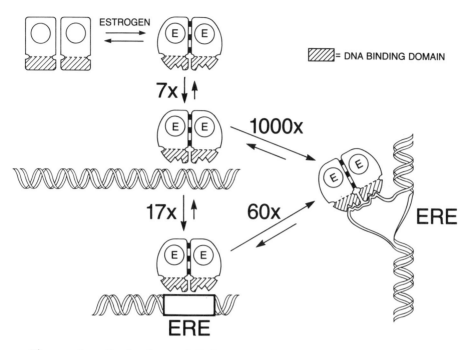

ERE

Figure 1. Mechanism of Estrogen Receptor Interaction with DNA (from Lannigan and Notides, 1988).

DNA-binding domain. Estradiol binds to the dimer in a positively cooperative binding mechanism causing a conformational change in the receptor so that its affinity for total DNA increases by 7-fold. The receptor has a 17-fold greater affinity for the double-stranded ERE compared to nonspecific DNA and therefore the double-stranded ERE is preferentially bound by the receptor. A high affinity site forms when the strands separate. Strand separation could be caused by transient strand opening and supercoiling. Sequences flanking the ERE or additional proteins could contribute to the strand separation and folding of the coding strand to form the high affinity site for estrogen binding. The receptor has a 60-fold greater affinity for binding to the "coding strand" high affinity site compared to the double-stranded ERE. The receptor then preferentially binds and stabilizes the high affinity site. The greater affinity of the receptor for single-stranded nonspecific DNA compared to double-stranded specific ERE is not inconsistent with the double-stranded ERE acting as a concentration site. The number of single-stranded nonspecific sites would be small because of a low probability that nonspecific DNA could exist in the single-stranded state without additional stabilizing factors. The formation of the estrogen receptor-"coding strand" ERE complex would be a crucial step in the transcriptional activation by estrogen.

Implicit in our findings is that DNA secondary structure plays a role in providing a high affinity binding site for the receptor. The formation of this site would not be a rate limiting step in estrogen regulation but would be an intrinsic property of the ERE, providing an additional level of molecular recognition for the receptor besides the ERE sequence.

This model of estrogen regulation of transcription relies on the steroid as the primary regulator. The positively cooperative binding mechanism of the steroid to the receptor enhances the receptor's responsiveness to small variations in the concentration of circulating steriod. The basis of selectivity that limits estrogen action to specific genes is the affinity of the receptor for its high affinity DNA binding site, which is controlled by the conformational change induced in the receptor by steroid binding.

REFERENCES

Auricchio F, Migliaccio A, Castoria G, Rotondi A, Di Domenico M, Pagano M, Nola E (1987). Phosphorylation on tyrosine of oestradiol-17ß receptor in uterus and interaction of oestradiol-17ß and glucocorticoid receptors with antiphosphotyrosine antibodies. J. Steroid Biochem. 27:245-253.
Burch JBE, Evans MI, Friedman TM, O'Malley PJ (1988). Two functional estrogen response elements are located upstream of the major chicken vitellogenin gene. Mol. Cell. Biol. 8:1123-1131.
Feavers IM, Jiricny J, Moncharmont B, Salvz HP, Jost JP (1987). Interaction of two nonhistone proteins with the estradiol response element of the avian vitellogenin gene modulates the binding of estradiol-receptor complex. Proc. Natl. Acad. Sci. USA 84:7453-7547.

Fried M, Crothers DM (1981). Equilibria and kinetics of lac repressor-operator interactions by polyacrylamide gel electrophoresis. Nucl. Acids Res. 9:6505-6525.

Gorski J, Toft D, Shyamala G, Smith D, Notides A (1968). Hormone Receptors: Studies on the interaction of estrogen with the uterus. In Astwood EB (ed): "Recent Progress in Hormone Research," New York: Academic Press, pp 45-80.

Hendrickson W (1985). Protein-DNA interactions studied by the gel electrophoresis-DNA binding assay. Biotechniques 3:198-207.

Jensen EV, DeSombre ER (1973). Estrogen-receptor interaction. Science 182:126-134.

Katzenellenbogen BS, Gorski J (1975). Estrogen actions on synthesis of macromolecules in target cells. In Litwack G (ed): "Biochemical Actions of Hormones, Vol. 3," New York pp 187-243.

Klein-Hitpass L, Ryffel G, Heitlinger E, Cato ACB (1988). A 13 bp palindrome is a functional estrogen responsive element and interacts specifically with estrogen receptor. Nucl. Acids Res. 16:647-663.

Klein-Hitpass L, Schorpp M, Wagner U, Ryffel GU (1986). An estrogen-responsive element derived from the 5' flanking region of the Xenopus vitellogenin A2 gene functions in transfected human cells. Cell 46:1053-1061.

Klock G, Strähle U, Schütz G (1987). Oestrogen and glucocorticoid responsive elements are closely related but distinct. Nature 329:734-736.

Kumar SA, Dickerman HW (1985). Steroid receptor-DNA interactions. In Moudgil VK (ed): "Molecular Mechanism of Steroid Hormone Action," New York: Walter de Gruyter, pp 505-538.

Lannigan DA, Notides AC (submitted). Estrogen receptor selectively binds the "coding strand" of an estrogen responsive element.

Martinez E, Givel F, Wahli W (1987). The estrogen-responsive element as an inducible enhancer: DNA sequence requirements and conversion to a glucocorticoid responsive element. EMBO 6:3719-3727.

Maurer RA (1985). Selective binding of the estradiol receptor to a region at least one kilobase upstream from the rat prolactin gene. DNA 4:1-9.

Maurer RA, Notides AC (1987). Identification of an estrogen-responsive element from the 5'-flanking region of the rat prolactin gene. Mol. Cell. Biol. 7:4247-4254.

Metzger D, White JH, Chambon P (1988). The human oestrogen receptor functions in yeast. Nature 334:31-36.

Nelson C, Crenshaw III EB, Franco R, Lira SA, Albert VR, Evans RM, Rosenfeld MG (1986). Discrete cis-active genomic sequences dictate the pituitary cell type-specific expression of rat prolactin and growth hormone genes. Nature 322:557-562.

Notides AC, Lerner N, Hamilton DE (1981). Positive cooperativity of the estrogen receptor. Proc. Natl. Acad. Sci. USA 78:4926-4930.

Notides AC, Nielsen S (1975). A molecular and kinetic analysis of estrogen receptor transformation. J. Steroid Biochem. 6:483-486.

Notides AC, Sasson S (1983). The positive cooperativity of the estrogen receptor and its relationship to receptor activation. In Eriksson H, Gustafsson JA (eds): "Steroid Hormone Receptors: Structure and Function," New York: Elsevier pp 103-120.

Notides AC, Sasson S, Callison S (1985). An allosteric regulatory mechanism for estrogen receptor activation. In Moudgil VK (ed): "Molecular Mechanism of Steroid Hormone Action," New York: Walter deGruyter, pp 279-299.

Notides, AC, Nielsen S (1974). The molecular mechanism of the in vitro 4S to 5S transformation of the uterine estrogen receptor. J. Biol. Chem. 249:1866-1873.

Redevilh G, Secco C, Mester J, Baulieu E (1987). Transformation of the 8-9S molybolate-stabilized estrogen receptor from low-affinity to high-affinity state without dissociation into subunits. J. Biol. Chem. 262:5530-5535.

Shupnik MA, Weinmann CM, Notides AC, Chin WW (in press). An upstream region of the rat LHß gene binds estrogen receptor and confers estrogen responsiveness. J. Biol. Chem.

Skafar DF, Notides AC (1987). The allosteric estrogen- and DNA-binding mechanism of the estrogen receptor. In Litwack G (ed): "Biochemical Actions of Hormones Vol. 14," New York: Academic Press, pp 317-346.

Skafar DF, Notides AC (1985). Modulation of the estrogen receptor's affinity for DNA by estradiol. J. Biol. Chem. 260:12208-12213.

Toft DO, Sullivan WP, McCormick DJ, Riehl RM (1987). Heat shock proteins and steroid hormone receptors. In Litwack G (ed): "Biochemical Actions of Hormones," New York: Academic Press, pp 293-316.

Waterman M, Adler S, Nelson C, Greene GL, Evans RM, Rosenfeld MG (1988) A single domain of the estrogen receptor confers deoxyribonucleic acid binding and transcriptional activation of the rat prolactin gene. Mol. Endocrin. 2:14-21.

IV. ESTROGEN ACTION AND BIOLOGICAL RESPONSES

Molecular Endocrinology and Steroid
Hormone Action, pages 201–211
© 1990 Alan R. Liss, Inc.

ESTROGEN REGULATION OF PROLIFERATION AND HORMONAL MODULATION
OF ESTROGEN AND PROGESTERONE RECEPTOR BIOSYNTHESIS AND
DEGRADATION IN TARGET CELLS

Benita S. Katzenellenbogen, Ann M. Nardulli
and Linnea D. Read
Department of Physiology and Biophysics, University
of Illinois and University of Illinois College of
Medicine, Urbana, Illinois 61801, USA

INTRODUCTION

Estrogenic hormones are known to stimulate a variety of
biosynthetic processes in hormone-responsive breast cancer
and uterine cells, and antiestrogens have been shown to
antagonize many of the actions of estrogens
(Katzenellenbogen et al., 1979, 1985; Aitken and Lippman,
1985). Indeed, antiestrogens have proven to be effective in
controlling the growth of estrogen-responsive breast cancers
(McGuire, 1979). The actions of estrogens appear to be
mediated via interaction with an intracellular receptor
protein (Sheridan et al. 1979; Welshons et al., 1984; King
and Greene, 1984). Ligand-free estrogen receptors are
weakly associated with nuclear components. Following ligand
binding, receptor complexes become tightly associated with
specific nuclear components and this association alters gene
expression (Katzenellenbogen, 1980; Gorski and Gannon, 1976;
Yamamoto, 1985). Antiestrogens also bind directly to the
estrogen receptor, and the resulting antiestrogen-receptor
complexes also become associated with chromatin, but appear
to block the events which promote cell growth
(Katzenellenbogen et al., 1984, 1985; Coezy et al., 1982).

Estrogen responsive breast cancer cell lines, and most
notably the MCF-7 human breast cancer cell line, have been
used extensively in studies aimed at analyzing the
mechanisms by which hormones affect cell proliferation and
protein synthesis. In MCF-7 cells, which contain functional
estrogen receptors, estrogen stimulates cell proliferation,
pS2 mRNA levels, plasminogen activator activity, thymidine
incorporation and DNA synthesis, and progesterone receptor
levels (Katzenellenbogen et al., 1985). Estrogen treatment
of MCF-7 cells also results in the stimulation of two
specific secreted proteins (M_r 160,000 and M_r 52,000) and a

cytoplasmic protein (M_r 24,000) (Westley et al., 1984; Edwards et al., 1981). In addition, we (Sheen and Katzenellenbogen, 1987) and others (Bronzert et al., 1987) have shown that antiestrogens specifically stimulate the production of an M_r 37,000 secreted glycoprotein, and that production of this protein is turned off by estrogen.

In order to gain further insight into how estrogens and antiestrogens modulate cell function, we have examined the effects of estrogen and antiestrogen on cell proliferation and on the synthesis of specific secreted proteins by breast cancer cells. We have also examined the factors regulating the levels of progesterone receptors and estrogen receptors in target cells.

EFFECTS OF SHORT-TERM AND LONG-TERM ESTROGEN WITHDRAWAL ON BREAST CANCER CELL PROLIFERATION AND GROWTH RESPONSIVENESS TO ESTROGEN AND ANTIESTROGEN

We undertook to compare the proliferation and estrogen and antiestrogen responsiveness of MCF-7 breast cancer cells grown in the short-term and long-term absence of estrogens. Since we found that components in phenol red preparations, added routinely to tissue culture media as a pH indicator, have weak estrogenic activity (Berthois et al., 1986; Bindal et al., 1988; Bindal and Katzenellenbogen, 1988), we grew cells in medium without phenol red and with charcoal dextran-treated serum to eliminate all known sources of estrogens for brief and for extended periods of time.

As seen in Figure 1, control cells grown in the presence of phenol red (Figure 1, left) proliferate at a rapid rate that is only slightly increased by estradiol. Hydroxytamoxifen decreased proliferation of these cells and hydroxytamoxifen inhibited the cell proliferation stimulated by 10^{-9}M estradiol, reducing the cell number to below that of the control.

MCF-7 cells exhibit a reduced growth rate when they are transferred to tissue culture medium that lacks phenol red. As seen in Figure 1 (middle bars), addition of estradiol (10^{-9}M) to cells grown in the absence of phenol red for several days (short-term phenol red-free cells) resulted in a marked increase in their proliferation rate,

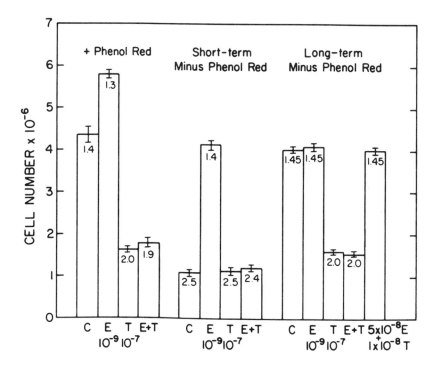

Figure 1. Proliferation rates of MCF-7 cells and the effect of estradiol (E) and trans-hydroxytamoxifen (T) on the proliferation of MCF-7 cells grown in the continuous presence of phenol red (left-most bars); or grown in the short-term (1 week) or long-term (6 months) absence of phenol red. Cells were seeded into T-25 flasks and, at two days after cell seeding, triplicate flasks of cells were counted and each medium was then supplemented with 10^{-9}M E, 10^{-7}M T, E + T, control ethanol vehicle (0.1%, C), or with 5 x 10^{-8}M E + 1 x 10^{-8}M T. Media and hormones were changed every other day and on day 7, triplicate flasks of cells were counted. Values represent the mean ± SEM of the values for each group and are representative of 3 separate experiments. The numbers inside each bar indicate the cell doubling time in days. (From Katzenellenbogen et al., 1987)

up to that seen in cells maintained in the presence of phenol red. Addition of the antiestrogen trans-hydroxytamoxifen (10^{-7}M) to these short-term phenol red-free cells had no effect on their proliferation rate, but antiestrogen did suppress the stimulatory effect of concomitant estradiol.

When MCF-7 cells were grown in the absence of phenol red for a period of 1 month prior to assay, their growth rate remained reduced, and their hormonal responsiveness to estradiol and trans-hydroxytamoxifen was similar to that seen in Figure 1, with short-term (ca. 1 week) phenol red withdrawn cells. The behavior of these short-term estrogen withdrawn cells is in contrast to that of cells maintained for much longer periods - five-six months - in the apparently complete absence of estrogens (no phenol red and with charcoal dextran-treated calf serum). These long-term estrogen-withdrawn cells (Figure 1, right bars) have an increased rate of proliferation, and the addition of estradiol causes no further increase in this maximal rate of proliferation, but antiestrogen decreases proliferation (Figure 1-right). The reduction of proliferation by trans-hydroxytamoxifen (10^{-8}M, which gave suppression equal to that seen with 10^{-7}M trans-hydroxytamoxifen) was reversed by estradiol (5 x 10^{-8}M, or 10^{-8}M).

Of great interest is our finding that these cells grown in the long-term absence of estrogen are now able to form tumors in nude mice in the absence of added estrogen. However, tumors occur at a higher incidence and are larger when estrogen is given (B.S. Katzenellenbogen, R. Clarke and M. Lippman, in preparation). Hence the cells are no longer fully estrogen-dependent for growth in vivo but they are still estrogen-responsive.

ESTROGEN RECEPTOR AND PROGESTERONE RECEPTOR LEVELS IN CELLS GROWN IN THE SHORT-TERM AND LONG-TERM ABSENCE OF ESTROGENS

While we found cells grown in the short-term absence of estrogens to have estrogen receptor levels similar to those of control cells, cells grown in the long-term absence of estrogens had 3-times higher estrogen receptor levels (Katzenellenbogen et al., 1987). As expected, progesterone receptor levels were very low in cells grown in the absence of phenol red for either a short time (5 days) or a long

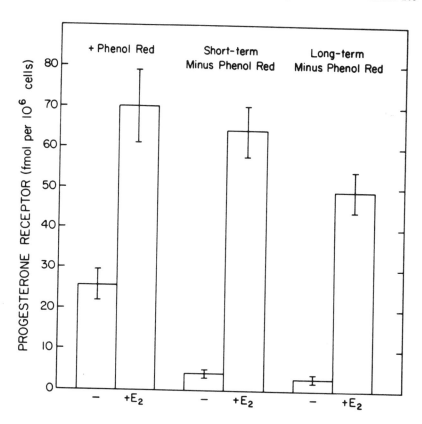

<u>Figure 2</u>. Progesterone receptor concentrations in control
and estradiol-treated MCF-7 cells grown in the continuous
presence of phenol red, or in the short-term (one week) or
long-term (six months) absence of phenol red. Progesterone
receptor levels in these different MCF-7 cells were
determined after five days of growth in the absence (-) or
presence (+) of 1 nM estradiol. Values indicate the mean ±
SEM of data obtained from three separate experiments. (From
Katzenellenbogen <u>et</u> <u>al</u>., 1987).

time (greater than 5-6 months). As seen in Figure 2, the control level of progesterone receptor in the short-term and long-term phenol red-free cells was significantly lower than that of cells grown with phenol red. In addition, the fold stimulation of progesterone receptor by estradiol was much greater in both the short-term and long-term phenol red-free cells than in control cells. Hence, both the long-term and short-term phenol red-free cells have low progesterone receptor levels that are markedly stimulated by estrogen, but they differ in their growth rate, with the long-term phenol red-free cells showing a rapid proliferation rate.

ESTROGEN-GROWTH FACTOR INTERRELATIONSHIPS

Our studies (Katzenellenbogen et al., 1987) indicate that MCF-7 cells respond almost immediately to the lack of estrogens with a decreased proliferation rate. In these short-term withdrawn cells, estradiol stimulated cell proliferation markedly while the antiestrogen trans-hydroxytamoxifen showed mixed partial agonist/antagonist activity. It weakly stimulated proliferation at low concentrations, but had no effect on proliferation at high concentrations. In addition, trans-hydroxytamoxifen showed a dose-dependent antagonism of estradiol-stimulated cell proliferation. The reduced rate of proliferation in the absence of estrogens is maintained by MCF-7 cells for at least one month, but by 5-6 months of growth in a phenol red-free environment, the "basal" proliferation rate of the cells has increased, returning to that of control cells maintained in continuous phenol-red. While estradiol was not able to further stimulate the proliferation of these long-term withdrawn cells, antiestrogen was still inhibitory to cell proliferation and estradiol was able to reverse the hydroxytamoxifen inhibition.

These studies document that the proliferation rate of MCF-7 cells is modulated markedly by the conditions under which the cells are grown, and that MCF-7 cells increase their basal growth rate in the long-term absence of estrogens. It is possible that this may represent selection of a subpopulation of MCF-7 cells during long-term culture in the total absence of estrogenic stimulation. This possibility is being investigated presently. The increased proliferation rate of cells grown in the long-term absence of estrogen does not appear to be due to the acquisition of supersensitivity to possible low levels of estrogen, since the basal progesterone receptor content of the long term-withdrawn cells is very low. Indeed, these cells remain highly sensitive to estrogen, since their progesterone receptor content is increased markedly by estradiol. In fact, their cellular estrogen receptor levels are increased approximately 3-fold. Welshons and Jordan (1987) have observed a similar increase in estrogen receptor levels in MCF-7 cells grown in the absence of phenol red.

It has been well documented that MCF-7 cells have receptors for many hormones and growth factors, and have their growth influenced by many of these agents (Dickson and Lippman, 1988). Recent studies have also shown that MCF-7 cells produce a variety of growth factors (TGF-α/EGF, TGF-β, IGF-1 and probably others) and that the production of some of these growth factors and growth inhibitors is modulated by estrogen and antiestrogen (Dickson and Lippman, 1988; Knabbe et al., 1987). Hence, it is possible that in the long-term absence of estrogen, there is either an altered production and/or secretion of growth factors or a change in sensitivity to growth factors or growth inhibitory factors present in the serum or secreted by the MCF-7 cells in culture. Studies now in progress are examining these aspects.

LIGAND MODULATED REGULATION OF PROGESTERONE RECEPTOR AND ESTROGEN RECEPTOR mRNA AND PROTEIN LEVELS

We have used dense amino acid labeling of receptors and cDNA probes to measure estrogen receptor and progesterone receptor mRNA levels, and to investigate receptor synthesis and degradation and their possible regulation by estrogen, progestin, antiestrogen, antiprogestin and by growth factors. These studies have revealed that the major effect of estrogen on increasing cellular progesterone receptor content is achieved by increasing the rate of progesterone receptor synthesis (Nardulli et al., 1988) and that this is reflected in increased levels of progesterone receptor mRNA (Nardulli et al., 1988; Read et al., 1988). Progestin evokes a time-and concentration-dependent decrease in progesterone receptor levels and our studies employing dense amino acids to distinguish newly synthesized and preexisting progesterone receptor reveal that the progestin evoked reduction in progesterone receptor levels is due both to a marked increase in the rate of progesterone receptor degradation as well as a dramatic decrease in the rate of receptor synthesis (Nardulli and Katzenellenbogen, 1988; Mullick and Katzenellenbogen, 1986). Reduction in all five species of PR mRNA accompany the decrease in progesterone receptor protein levels (Read et al., 1988; Wei et al., 1988).

Estrogen receptor mRNA and protein levels in MCF-7 cells are decreased on exposure to estradiol and this down-regulation by estrogen, which is estrogen concentration-dependent, is antagonized by concomitant antiestrogen treatment. Progestins also modulate estrogen receptor levels in MCF-7 and T47D breast cancer cells. However, the growth factors EGF, IGF1 and TGF-β have very little effect on estrogen receptor levels, suggesting that the steroid hormone ligands rather than growth factors are the major regulators of the estrogen receptor content of these breast cancer cells (Read and Katzenellenbogen, 1988).

ACKNOWLEDGEMENTS

We are grateful for support of this research by NIH grants CA 18119 and HD 21524 (to B. S. K.).

REFERENCES

Aitken SC, Lippman ME (1985). Effect of estrogens and antiestrogens on growth regulatory enzymes in human breast cancer cells in tissue culture. Cancer Res 45:1611-1620.

Berthois Y, Katzenellenbogen JA, Katzenellenbogen BS (1986). Phenol red in tissue culture media is a weak estrogen: implications concerning the study of estrogen-responsive cells in culture. Proc Natl Acad Sci USA 83:2496-2500.

Bindal RD, Carlson KC, Norman MJ, Katzenellenbogen BS, Katzenellenbogen JA (1988). Lipophilic impurities, not phenosulfonphthalein, account for the estrogenic activity in commercial preparations of phenol red. J Steroid Biochem, in press.

Bindal RD, Katzenellenbogen JA (1988). Bis- (4-hydroxyphenyl)-[2-(phenoxysulfonyl)phenyl] methane: isolation and structure elucidation of a novel estrogen from commercial preparations of phenol red (phenolsulfonphtalein) J Med Chem, in press.

Bronzert DA, Silverman S, Lippman ME (1987). Estrogen inhibition of a M$_r$ 39,000 glycoprotein secreted by human breast cancer cells. Cancer Res. 47:1234-1238.

Coezy E, Borgna JL, Rochefort H (1982). Tamoxifen and metabolites in MCF-7 cells: correlation between binding to estrogen receptor and inhibition of cell growth. Cancer Res 42:317-324.

Dickson RB, Lippman ME (1987). Estrogenic regulation of growth and polypeptide growth factor secretion in human breast carcinoma. Endocrine Reviews 8:29-42.

Edwards DP, Adams DJ, McGuire WL (1981). Estradiol stimulates synthesis of a major intracellular protein in a human breast cancer cell line (MCF-7). Breast Cancer Res Treatment 1:209-215.

Gorski J, Gannon F (1976). Current models of steroid hormone action: A critique. Ann Review Physiol 38:425-450.

Katzenellenbogen BS (1980). Dynamics of steroid hormone receptor action. Ann Review Physiol 42:17-35.

Katzenellenbogen BS, Bhakoo HS, Ferguson ER, Lan NC, Tatee T, Tsai TL, Katzenellenbogen, JA (1979). Estrogen and antiestrogen action in reproductive tissues and tumors. Recent Prog Hormone Res 35:259-300.

Katzenellenbogen BS, Kendra KL, Norman MJ, Berthois Y (1987). Proliferation, hormonal responsiveness, and estrogen receptor content of MCF-7 human breast cancer cells grown in the short-term and long-term absence of estrogens. Cancer Res 47:4355-4360.

Katzenellenbogen BS, Miller MA, Mullick A, Sheen YY (1985). Antiestrogen action in breast cancer cells: modulation of proliferation and protein synthesis, and interaction with estrogen receptors and additional antiestrogen binding sites. Breast Cancer Res Treatment 5:231-243.

Katzenellenbogen BS, Norman MJ, Eckert RL, Peltz SW, Mangel WF (1984). Bioactivities, estrogen receptor interactions and plasminogen activator inducing activities of tamoxifen and hydroxytamoxifen isomers in MCF-7 human breast cancer cells. Cancer Res 44:112-119.

King WJ, Greene GL (1984). Monoclonal antibodies localize estrogen receptor in the nuclei of target cells. Nature 307:745-747.

Knabbe C, Lippman ME, Wakefield LM, Flanders KC, Kasid A, Derynck R, Dickson RB (1987). Evidence that transforming growth factor-β is a hormonally regulated negative growth factor in human breast cancer cells. Cell 48:417-428.

McGuire WL (1979). Steroid receptor sites in cancer therapy. Adv Intern Med 24:127-140.

Mullick A, Katzenellenbogen BS (1986). Progesterone receptor synthesis and degradation in MCF-7 human breast cancer cells as studied by dense amino acid incorporation: evidence for a non-hormone binding receptor precursor. J Biol Chem 261:13236-13243.

Nardulli AM, Greene GL, O'Malley BW, Katzenellenbogen BS (1988). Regulation of progesterone receptor messenger ribonucleic acid and protein levels in MCF-7 cells by estradiol: Analysis of estrogen's effect on progesterone receptor synthesis and degradation. Endocrinology 122:935-944.

Nardulli AM, Katzenellenbogen BS (1988). Progesterone receptor regulation in T47D human breast cancer cells: Analysis by density labeling of progesterone receptor synthesis and degradation and their modulation by progestin. Endocrinology 122:1532-1540.

Read LD, Katzenellenbogen BS (1988). Regulation of estrogen receptor mRNA and protein levels in human breast cancer cell lines by sex steroid hormones, their antagonists and growth factors. Endocrinology 122 (Suppl):129.

Read LD, Snider CE, Miller JS, Greene GL, Katzenellenbogen BS (1988). Ligand-modulated regulation of progesterone receptor messenger ribonucleic acid and protein in human breast cancer cell lines. Molec Endocrinology 2:263-271.

Sheen YY, Katzenellenbogen BS (1987). Antiestrogen stimulation of the production of a 37,000 molecular weight secreted protein and estrogen stimulation of the production of a 32,000 molecular weight secreted protein in MCF-7 human breast cancer cells. Endocrinology 120:1140-1151.

Sheridan PJ, Buchanan JM, Anselomo VC, Martin PM (1979). Equilibrium: The intracellular distribution of steroid receptors. Nature 282:579-584.

Wei LL, Krett NL, Francis MD, Gordon DF, Wood WM, O'Malley BW, Horwitz KB (1988). Multiple human progesterone receptor messenger ribonucleic acids and their autoregulation by progestin agonists and antagonists in breast cancer cells. Molec Endocrinology 2:62-68.

Welshons WV, Jordan VC (1987). Adaptation of estrogen-dependent MCF-7 cells to low estrogen (phenol red-free) culture. Eur J Cancer Clin Oncol. 23:1935-1939.

Welshons WV, Lieberman ME, Gorski J (1984). Nuclear localization of unoccupied estrogen receptors. Nature 307:747-749.

Westley B, May EB, Brown AM, Krust A, Chambon P, Lippman ME, Rochefort H (1984). Effects of anitestrogens on the estrogen regulated pS2 RNA and 52 and 160 kilodalton proteins in MCF-7 cells and two tamoxifen resistant sublines. J Biol Chem 259:10030-10035.

Yamamoto KR (1985). Steroid receptor regulated transcription of specific genes and gene networks. Ann Rev Genetics 19:209-252.

Molecular Endocrinology and Steroid
Hormone Action, pages 213–226
© 1990 Alan R. Liss, Inc.

REGULATION OF THE UTERINE EPIDERMAL GROWTH FACTOR RECEPTOR BY ESTROGEN

G.M. Stancel[*], C. Chiapetta[*], R.M. Gardner[*], J.L. Kirkland[+], T.H. Lin[+], R.B. Lingham[*], D.S. Loose-Mitchell[*], V.R. Mukku[*], and C.A. Orengo[*]
*Dept. of Pharmacology, University of Texas Medical School, Houston, Texas, 77225, and +Div. of Endocrinology, Dept. of Pediatrics, Baylor College of Medicine, Houston, Texas 77030

INTRODUCTION

Epidermal growth factor is a polypeptide isolated originally by Cohen on the basis of its ability to accelerate eyelid opening and incisor eruption in mice (Cohen, 1962). Since that time this growth factor has been extensively studied and found to have growth promoting effects in a large number of systems (Carpenter and Cohen, 1979). It has also been recognized for some time that EGF and/or EGF-like peptides may play a role in paracrine and autocrine mechanisms of cellular growth control (Carpenter, 1981). With this background, a number of laboratories interested in the regulation of growth by steroid hormones have examined a possible role for EGF in the growth of estrogen sensitive tissues and cells.

One such system is the rodent uterus, which has served historically as a major model for studying the effects of estrogens on the growth of normal target tissues. Various studies have demonstrated that: 1) the uterus and uterine luminal fluid contain EGF like material (Gonzalez et al., 1984; Imai, 1982; DiAugustine et al., 1985; Diaugustine et al., 1988); 2) prepro-EGF mRNA is found in the uterus (DiAugustine et al.,

1988); 3) EGF stimulates the growth of uterine epithelial cells (Tomooka et al., 1986) and cells derived from uterine smooth muscle (Bhargava et al., 1979); and, 4) the uterus contains EGF receptors (Mukku and Stancel, 1985a; Mukku and Stancel, 1985b; Hofmann et al., 1984) and EGF receptor mRNA (Lingham et al., 1988). An especially important observation has been made by McLachlan et al. (1987) who reported recently that antibodies against EGF block the estrogen sensitive growth of uterine cells in organ culture. Taken together these observations support a role for EGF in the physiological control of uterine growth by estrogens.

Given this possibility we decided to investigate the structure, function and regulation of EGF receptors in the uterus. Receptor levels are one obvious potential determinant of tissue responsiveness to growth factors, and receptor levels may be especially important for ligands like EGF which "down regulate" membrane receptors. For example, it has been known for some time that a brief, transient exposure of cells to EGF is not sufficient to stimulate DNA replication (Carpenter and Cohen, 1979), and Knauer et al. (1984) more recently illustrated that there is a direct relationship between EGF receptor occupancy and the mitogenic response of fibroblasts in a steady state model. In this chapter we thus describe our studies on the properties of the uterine EGF receptor and its regulation by estrogen.

PROPERTIES OF THE UTERINE EGF RECEPTOR

EGF produces its effects by interacting with specific, high-affinity receptors present in target cells (Carpenter, 1987). As seen in Figure 1, membranes prepared from the immature rat uterus contain saturable, high-affinity EGF binding sites. In a number of studies using different species (mouse and rat), different age animals (immature and mature), and different endocrine states we have observed that uterine membranes contain only a single class of EGF binding sites

with a Kd value in the range of 0.4 - 2 nM EGF.
Membranes prepared from animals not exposed to
estrogen generally contain about 200 fmoles of EGF
binding sites per mg of membrane protein.
Estrogen treatment in vivo increases the number of
functional EGF receptors (see below), but does not
alter the affinity of the receptor for the growth
factor (Mukku and Stancel, 1985b). These binding
sites are specific for EGF since the binding of
^{125}I-EGF is displaced by an excess of the
unlabelled growth factor but not by other peptides
(Mukku and Stancel, 1985a).

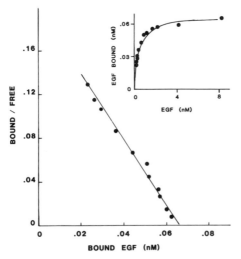

Figure 1.   Specific binding of ^{125}I-EGF to
isolated uterine membranes. Reprinted from Mukku
and Stancel (1985a) with permission.

        In the immature rat uterus, EGF binding sites
are observed in luminal and glandular epithelium,
stroma and myometrium using labelled ligand
binding to tissue segments followed by
autoradiography. A number of control studies
revealed that these binding sites represent
authentic EGF receptors (Lin et al., 1988). It is
important to note that in the immature animal all
these cell types undergo a growth response

following estrogen administration in vivo (Kaye et al., 1972). Similarly, Chegini et al. (Chegini et al., 1986) have used autoradiography to determine that all major cell types of the human uterus contain EGF receptors, and several groups have used ligand binding to illustrate that endometrial (Tomooka et al., 1986; Hofmann et al., 1984) and myometrial (Bhargava et al., 1979; Hofmann et al., 1984) cells contain EGF receptors.

The mechanism of signal transduction of the EGF receptor appears to be activation of a tyrosine protein kinase activity upon ligand binding (Carpenter, 1987). We have shown that incubation of solubilized uterine membrane preparations with EGF stimulates autophosphorylation of the 170,000 MW EGF receptor (Mukku and Stancel, 1985a and b). Additional studies have shown that this phosphorylation occurs primarily on tyrosine residues of the receptor (Mukku and Stancel, 1985a), and that EGF binding also stimulates kinase activity measured with an exogenous substrate containing a tyrosine residue (Mukku and Stancel, 1985b). Chemical crosslinking studies with the bifunctional reagent dissuccinimidyl-suberate have established also that the molecular weight of the uterine EGF receptor is 170,000 (Mukku and Stancel, 1985 a and b). Taken together these results illustrate that the uterine EGF receptor is similar to that reported in a wide variety of other tissues and cells (Carpenter, 1987).

REGULATION OF THE UTERINE EGF RECEPTOR BY ESTROGEN

Administration of estradiol to immature animals produces a 2 - 3 fold increase in functional EGF receptors assessed by ligand binding (Figure 2). This effect occurs primarily between 6 and 12 hours after hormone treatment, and is specific for estrogenic steroids (Mukku and Stancel, 1985b). This increase occurs well before tissue DNA synthesis which begins to increase roughly 15 hours after steroid administration and is not maximum until 21 -24 hours after hormone

treatment (Kaye et al., 1972). Estradiol
treatment also produces a comparable increase in
tyrosine kinase activity of the EGF receptor
(Mukku and Stancel, 1985b).

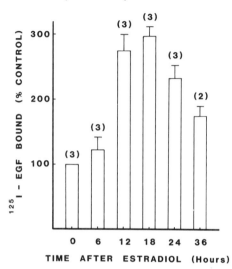

Figure 2. Regulation of uterine EGF receptors by
estrogen. Immature rats were treated with
estradiol for the indicated times prior to
sacrifice. Uterine membranes were prepared and
analyzed for specific ^{125}I-EGF binding. Taken
from Mukku and Stancel (1985b) with permission.

The studies illustrated in Figure 2 were
performed with immature rats. More recent studies
have established that estrogen increases uterine
EGF receptor levels in the immature mouse (Gardner
et al., in preparation) and in the castrate adult
rat (Gardner et al., submitted). In addition, the
level of uterine EGF receptor varies throughout
the estrous cycle in rats in parallel with changes
in plasma estrogens and occupied tissue estrogen
receptors (Gardner et al., submitted). The
generality of this effect suggests that the
regulation of uterine EGF receptors by estrogen is
a physiological effect.

The induction of uterine EGF receptors by estrogen is sensitive to both cycloheximide and actinomycin D (Mukku and Stancel, 1985b) suggesting that the observed increases represent de novo receptor synthesis and that the mechanism of induction is at least partially transcriptional in nature. This possibility received further support from the demonstration that estrogen treatment in vivo increases the level of EGF receptor mRNA (Lingham et al., 1988). As illustrated in Figure 3 estradiol treatment of immature rats leads to an increase in the 9.5 Kb EGF receptor transcript prior to an increase in functional receptor levels. This increase in the EGF receptor mRNA is sensitive to actinomycin D but not puromycin, and the increase is specific for estrogenic steroids (Lingham et al., 1988). We have also observed a similar increase in the mouse uterine EGF receptor mRNA following estradiol treatment in vivo (Gardner et al., unpublished observation).

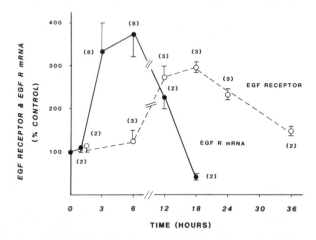

Figure 3. Regulation of EGF receptor mRNA by estrogen. Immature rats were treated with estradiol for the indicated times prior to sacrifice. mRNA levels were determined by densitometric analysis of RNA blots hybridized with a riboprobe complementary to the rat EGF receptor cDNA. EGF receptor levels were determined by ligand binding. Reprinted from Lingham et al. (1988) with permission.

One interpretation of these results is that the induction of the EGF receptor is a primary effect of estradiol acting via its nuclear receptor. This mechanism is compatible with our existing data, but other experiments such as direct measurements of transcription rates, evaluation of message stability and a search for estrogen responsive elements in the EGF receptor gene region will be required to unequivocally establish this point.

It is also conceivable that estrogen might act initially through a non-transcriptional mechanism to trigger the release of EGF from precursor sites (DiAugustine et al., 1988), and the EGF thus formed might be the stimulus to induce the production of its own receptor (Earp et al., 1986; Clark et al., 1985; Kudlow et al., 1986). While we believe this possibility is less likely, it cannot be ruled out on the basis of the available data.

RELATIONSHIP BETWEEN EGF RECEPTOR INDUCTION AND TISSUE DNA SYNTHESIS

We are currently investigating the relationship between the increases in uterine EGF receptor levels and DNA synthesis following estrogen treatment. While these studies are not yet complete, available results include the following: 1) dose response and hormonal specificity profiles for the two parameters are virtually identical; 2) a single injection of short acting estrogens does not appreciably increase receptor levels or DNA synthesis, but repeated administration of these compounds increases both to the same degree as estradiol treatment; and 3) increases in both parameters are diminished in the uterus to the same degree in hypothyroid animals (see Gardner et al., 1978, and Mukku, 1984, for background information). At present it appears that increases in tissue DNA synthesis after estrogen treatment correlate with, and are preceded by, increases in EGF receptor levels, although the evidence available does not

necessarily indicate a direct cause-effect
relationship.

One can envision two general ways that an
elevation in growth factor receptor synthesis
might play an important role in tissue DNA
synthesis. Increases in the EGF receptor level,
alone or in combination with increases in EGF or
EGF-like ligands, might be necessary to produce a
threshold level of a cellular second messenger
required for progression towards, or initiation
of, tissue DNA synthesis. Alternatively, cells
stimulated to grow might require an increased
synthesis of receptors to prevent down regulation
below the point necessary to sustain a prolonged
signal required for DNA synthesis. Further
studies will clearly be required to determine the
role of EGF receptor levels in estrogen stimulated
growth.

STIMULATION OF MYOMETRIAL CONTRACTIONS BY EGF

Most studies of EGF have focused on the
growth promoting effects of this peptide.
However, it is clear that this peptide also has
several other biological activities including the
ability to decrease gastric acid secretion
(Carpenter and Cohen, 1979; Gregory, 1985) and to
produce contractions of vascular smooth muscle
(Muramatsu et al., 1985; Berk et al., 1985).
Therefore, we investigated the possibility that
EGF might also stimulate myometrial contractions.

The addition of EGF to segments of uterine
tissue in an in vitro organ bath system rapidly
stimulates contractile activity (Figure 4). This
effect is produced by low concentrations of EGF
(ED_{50} of 3.5 nM), and is specific since it is not
produced by other peptides such as insulin or MSA
(Gardner et al., 1987). This response to EGF
requires in vivo estrogen priming of the tissue,
but it seems unlikely that this requirement is due
solely to the induction of growth factor receptor
by the steroid (Gardner et al., 1987). Other
studies have revealed that EGF stimulates the

myometrium directly, since contractions occur if
the endometrium is physically removed from the
muscle layer (Gardner and Stancel, submitted).

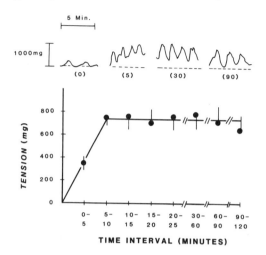

Figure 4. Stimulation of uterine contractions by
EGF. (Top) EGF was added _in vitro_ to an isolated
uterine segment from a mature castrate animal
which received estradiol priming _in vivo_ for 24
hours prior to sacrifice. The tracings represent
contractile activity at the indicated times after
EGF addition. (Bottom) Composite showing the
maximum tension developed _in vitro_ in the
indicated time intervals after EGF addition to a
series of isolated uterine tissues; N = 6-7 per
point. Taken from Gardner _et al_. (1987) with
permission.

While not yet complete, studies in progress
suggest that EGF causes the release of arachidonic
acid from uterine membrane sites. The arachidonic
acid then appears to be converted to both
prostaglandins and leukotrienes, which are well
known stimulators of myometrial contractility.
This possible mechanism for the production of
uterine contractions by EGF is tentative, however,
since it is based on the use of pharmacological
inhibitors rather than direct measurements of

arachidonic acid metabolites (Gardner and Stancel, submitted). It is clear nevertheless that EGF is a potent and efficacious stimulant of myometrial contractions, and this raises the clear possibility that EGF may produce physiological effects on the myometrium other than growth.

OTHER GROWTH FACTORS AND PROTOONCOGENES

It is becoming increasingly clear that normal uterine tissue contains a number of growth factors and receptors besides EGF. These include IGF-1 (Murphy et al., 1987), PDGF receptors (Ronnstrand et al., 1987) and UDGF, a uterine derived growth factor from sheep uterus (Ikeda and Sirbasku, 1984). It is especially important to note that in at least one case, IGF-1 has been shown to be regulated by estrogen treatment in vivo (Murphy et al., 1987). It should be noted that estrogen also regulates the production of IGF-1 and EGF-like peptides in estrogen sensitive MCF-7 breast cancer cells (Kasid and Lippman, 1987).

Similarly, it is apparent that estrogens stimulate the expression of various protooncogenes in the normal uterus. The EGF receptor itself is the cellular homolog of the erb B oncogene (Downward et al., 1984). Other groups have made the important observations that estrogen treatment increases the mRNA levels of c-myc (Travers and Knowler, 1987; Murphy et al., 1987), N-myc (Murphy et al., 1987) and c-Ha^{ras} (Travers and Knowler, 1987), and we have recently shown that estradiol increases uterine levels of c-fos mRNA (Loose-Mitchell et al., 1988) Estrogens thus appear to stimulate the expression of three general types of protooncogenes: tyrosine kinases such as the EGF receptor; nuclear protooncogenes such as c-myc,N-myc and c-fos; and analogues of cellular G-proteins such as c-Ha^{ras}.

Taken together these observations suggest that the regulation of normal uterine growth by estrogens may involve an interplay between the steroid hormone and a variety of growth factors

and protooncogene products. In addition these
findings may provide important clues to the
pathophysiology of aberrant growth of estrogen
target tissues. Finally, these studies suggest
that the uterus may be a system well suited for
studying the role of growth factors and
protooncogenes in normal tissue.

ACKNOWLEDGMENTS

We thank Ms. Lori Ezzell for preparation of
this manuscript. Research in our laboratories has
been supported by NIH grants HD-08615, DK-38965,
RR-01685 (Bionet) and by the John P. McGovern
Foundation. Current address: RBL - Merck and
Co., 80Y-310, Rahway, NJ, 07065; RMG - Dept.
Biology, Villanova Univ., Villanova, PA, 19085;
VRM - Genentech, Inc., 460 Point San Bruno Blvd.,
South San Francisco, CA, 94080.

REFERENCES

Berk BC, Brock TA, Webb RC, Taubman MB, Atkinson
    WJ, Gimbrone MA Jr, Alexander RW (1985).
    Epidermal growth factor, a vascular smooth
    muscle mitogen, induces rat aortic contraction.
    J Clin Invest 75:1083-1086.
Bhargava G, Rifas L, Makman MH (1979). Presence
    of epidermal growth factor receptors and
    influence of epidermal growth factor on
    proliferation and aging in cultured smooth
    muscle cells. J Cell Physiol 100:365-374.
Carpenter G (1981). Epidermal growth factor. In
    Baserga R (ed): "Tissue Growth Factors,"
    Berlin: Springer-Verlag, pp 89-132.
Carpenter G (1987). Receptors for epidermal
    growth factor and other polypeptide mitogens.
    Annu Rev Biochem 56:881-914.
Carpenter G, Cohen S (1979). Epidermal growth
    factor. Annu Rev Biochem 48:193-216.
Chegini N, Rao CV, Barrows GH, Sanfilippo JS
    (1986). Binding of ^{125}I-epidermal growth factor
    in human uterus. Cell Tissue Res 246:543-548.

Clark AJL, Ishii S, Richert N, Merlino GT, Pastan
I (1985). Epidermal growth factor regulates the
expression of its own receptor. Proc Natl Acad
Sci USA 82:8374-8378.
Cohen S (1962). Isolation of a mouse submaxillary
protein accelerating incisor eruption and eyelid
opening in the newborn animal. J Biol Chem
237:1555-1562.
DiAugustine RP, Lammon DE, McLachlan JA (1985).
Sex steroid hormones rapidly increase uterine
epidermal growth factor (EGF). Endocrinology
116(suppl):169 (abstr).
DiAugustine RP, Petrusz P, Bell GI, Brown CF,
Konach KS, McLachlan JA, Teng CT (1988).
Influence of estrogens on mouse uterine
epidermal growth factor precursor protein and
messenger RNA. Endocrinology 122:2355-2363.
Downward J, Yarden Y, Mayes D, Scrace G, Totty N,
Stockwell P, Ullrich A, Schlessinger J,
Waterfield MD (1984). Close similarity of
epidermal growth factor receptor and v-erb-B
oncogene protein sequences. Nature 307:521-527.
Earp HS, Austin KS, Blaisdell J, Rubin RA, Nelson
KG, Lee LG, Grishan JE (1986). Epidermal growth
factor (EGF) stimulates EGF receptor synthesis.
J Biol Chem 261:4777-4780.
Gardner RM, Kirkland JL, Ireland JS, Stancel GM
(1978). Regulation of the uterine response to
estrogen by thyroid hormone. Endocrinology
103:1164-1172.
Gardner RM, Lingham RB, Stancel GM (1987).
Contractions of the isolated uterus stimulated
by epidermal growth factor. FASEB J 1:224-228.
Gonzalez F, Lakshmanan J, Hoath S, Fisher DA
(1984). Effect of oestradiol-17B on uterine
epidermal growth factor concentration in
immature mice. Acta Endocrinol 105:425-428.
Gregory H (1985). In vivo aspects of urogastrone-
epidermal growth factor. J Cell Sci Suppl 3:11-
17.
Hofmann GE, Rao CV, Barrows GH, Sanfilippo JS
(1984). Binding sites for epidermal growth
factor in human uterine tissues and leiomyomas.
J Clin Endocrinol Metab 58:880-884.

Ikeda T, Sirbasku DA (1984). Purification and properties of a mammary-uterine-pituitary tumor cell growth factor from pregnant sheep uterus. J Biol Chem 259:4049-4964.

Imai Y (1982). Epidermal growth factor in rat uterine luminal fluid. Endocrinology 110 (suppl):162 (abstr).

Kasid A, Lippman ME (1987). Estrogen and oncogene mediated growth regulation of human breast cancer cells. J Steroid Biochem 27:465-470.

Kaye AM, Sheratzky D, Lindner HR (1972). Kinetics of DNA synthesis in immature rat uterus: age dependence and estradiol stimulation. Biochem Biophys Acta 261:475-486.

Knauer DJ, Wiley HS, Cunningham DD (1984). Relationship between EGF receptor occupancy and mitogenic response. J Biol Chem 259:5623-5631.

Kudlow JE, Cheung CYM, Bjorge JD (1986). Epidermal growth factor stimulates the synthesis of its own receptor in human breast cancer cell line. J Biol Chem 261:4134-4138.

Lin TH, Mukku VR, Verner G, Kirkland JL, Stancel GM (1988). Autoradiographic localization of epidermal growth factor receptors to all major uterine cell types. Biol Repro 38:403-411.

Lingham RB, Stancel GM, Loose-Mitchell DS (1988). Estrogen regulation of epidermal growth factor receptor messenger RNA. Mol Endocrinol 2:230-235.

Loose-Mitchell DS, Chiapetta C, Stancel GM (1988). Estrogen regulation of c-fos messenger RNA. Molec Endocrinol (in press).

McLachlan JA, DiAugustine RP, Newbold RR (1987). Estrogen induced uterine cell proliferation in organ culture is inhibited by antibodies to epidermal growth factor. Program of the 69th meeting of the Endocrine Society, Indianapolis, Abstr. 313, p 99.

Mukku VR (1984). Regulation of EGF receptor levels by thyroid hormone. J Biol Chem 259:6543-6547.

Mukku VR, Stancel GM (1985a). Receptors for epidermal growth factor in the rat uterus. Endocrinology 117:149-154.

Mukku VR, Stancel GM (1985b). Regulation of epidermal growth factor receptor by estrogen. J Biol Chem 260:9820-9824.

Muramatsu I, Hollenberg MD, Lederis K (1985). Vascular actions of epidermal growth factor-urogastrone: possible relationship to prostaglandin production. Can J Physiol Pharmacol. 63:994-999.

Murphy CJ, Murphy CC, Friesen HG (1987). Estrogen induces insulin-like growth factor-1 expression in the uterus. Molec Endocrinol 1:445-450.

Murphy LJ, Murphy LC, Friesen HG (1987). Estrogen induction of N-myc and c-myc protooncogene expression in the rat uterus. Endocrinology 120:1882-1888.

Ronnstrand L, Beckmann MP, Faulders B, Ostman A, Ek B, Heldin CH (1987). Purification of the receptor for PDGF from porcine uterus. J Biol Chem 262:2929-2932.

Tomooka Y, DiAugustine RP, McLachlan JA (1986). Proliferation of mouse uterine epithelial cells in vitro. Endocrinology 118:1011-1018.

Travers MT, Knowler JT (1987). Oestrogen induced expression of oncogenes in the immature rat uterus. FEBS Lett 211:27-30.

Molecular Endocrinology and Steroid
Hormone Action, pages 227–240
© 1990 Alan R. Liss, Inc.

ESTROGEN-INDUCED DESTABILIZATION AND ASSOCIATED DEGRADATION INTERMEDIATES OF APOLIPOPROTEIN II mRNA

Roberta Binder, David A. Gordon, Sheng-Ping L. Hwang, and David L. Williams

Department of Pharmacological Sciences, Health Sciences Center, State University of New York at Stony Brook, Stony Brook, New York 11794

INTRODUCTION

Numerous studies have demonstrated that mRNA stability plays an important role in gene regulation (Brawerman, 1987). This is the case with genes which respond to estrogenic steroids by both increased transcription and mRNA stabilization (Brock and Shapiro, 1983a and b; Shapiro et al., 1987). Originally described for ovalbumin mRNA in the chick oviduct (Palmiter and Carey, 1974), estrogen was also shown to induce the accumulation of conalbumin, ovamucoid, and lysozyme mRNAs in chick oviduct (Hynes et al., 1979), vitellogenin (VTG) II and apolipoprotein (apo) II mRNAs in chick liver (Wiskocil et al., 1980), and VTG mRNA in Xenopus liver (Shapiro and Baker, 1977). These mRNAs were believed to be stabilized by estrogen because hormone removal resulted in their rapid degradation. Workers in the field have proposed that the interaction of cellular factors with sequence or structural elements in the message may explain the hormone induced alteration in message stability.

Previously, we have shown that estrogen induces a destabilization activity which selectively degrades apoII and VTG II mRNAs upon hormone withdrawal (Gordon et al., 1988). The turnovers of chick liver apoII and VTG II mRNAs were measured in the presence of estrogen during the approach to a new steady state. Turnover was also measured following different lengths of estrogen treatment and subsequent hormone withdrawal. ApoII and VTG II mRNAs were selectively destabilized upon hormone withdrawal, but only after prolonged estrogen treatment (at least 5 days). During the period required to attain the new steady state level of these mRNAs (2 to 3 days), message half life was identical in the presence or absence of estrogen. Our group is using several approaches to determine the mechanism for estrogen induced destabilization of apoII mRNA upon hormone withdrawal. Site specific degradation intermediates of apoII mRNA have been identified both *in vivo* and in liver homogenates. These occur in the 3' non-coding region predominantly at or near the trinucleotide 5'-AAU-3' within predicted loop regions on a secondary structure model. The presence of both the 5' and 3' cleavage products suggests that these intermediates are produced by an endonucleolytic mechanism. Cleavage at these sites may be the rate limiting targeting event that leads to rapid destabilization of apoII mRNA during hormone withdrawal. Experimental evidence supporting these conclusions are as follows.

ESTROGEN-INDUCED DESTABILIZATION OF APOII AND VTGII mRNAs

ApoII and VTG II mRNA half lives were measured indirectly by using the approach to steady state method of Berlin and Schimke (1965) as modified by Watson et al. (1981). Chickens were sacrificed at intervals following implantation of constant release 17 β-estradiol pellets. Liver RNA was assayed for the presence of apoII mRNA by solution hybridization and for VTG II mRNA by dot blot analysis. The induction curves for these messages are shown in Fig. 1. Both curves show similar kinetics; when the data were plotted according to Watson et al. (1981), the accumulations of apoII and VTG II messages were described by one line that indicated a $t_{1/2}$ of 12.6 hours (Gordon et al., 1988).

ApoII and VTG II mRNA half lives were measured directly by removing hormone pellets and administering the anti-estrogen, tamoxifen citrate (Gordon et al., 1988). Liver RNA was assayed at various times following withdrawal. Fig. 2A plots the decay of apoII and VTG II mRNAs following 14 days of estrogen treatment. The $t_{1/2}$ of 1.5 hours for both messages demonstrates their rapid decay once hormone is removed. These results along with those in Fig. 1 support a model in which estrogen stabilizes estrogen-induced messages. However, as shown in Fig. 2B, apoII and VTG II mRNAs have a much slower turnover upon estrogen withdrawal following a 24 hour estrogen treatment ($t_{1/2}$=13 hours). This $t_{1/2}$ is the same as determined in the presence of estrogen from the approach to steady state (Fig. 1A and 1B). Therefore, it appears that estrogen regulates a process that alters the degradation of these mRNAs once the hormone is withdrawn. These data also indicate that estrogen does not alter the half life of apoII and VTG II mRNAs during the attainment of the steady state; the extensive accumulations of apoII and VTG II mRNAs appear to be due entirely to transcriptional activation of these genes. mRNA stabilization does not contribute to the accumulations of these mRNAs.

To determine when in the course of estrogen treatment the $t_{1/2}$ changes from a slow to rapid decay mode, hormone was withdrawn at various times following pellet implantation. ApoII and VTG II mRNAs were measured 3 hours post withdrawal. Fig. 3 shows that the change in turnover upon hormone withdrawal begins between

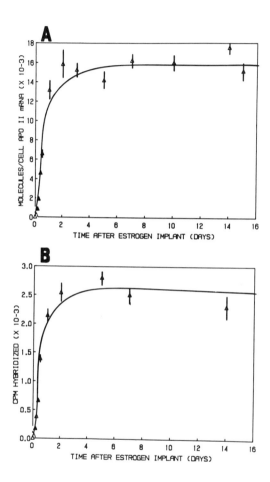

Figure 1. Approach to steady state of apoII and VTG II mRNAs. Male leghorn chickens, 2-4 weeks old, were treated with estrogen using subcutaneous implants containing 100 mg of 17 β-estradiol and were sacrificed at the indicated times. Liver RNA samples were assayed for apoII mRNA (A) and VTG II mRNA (B) as described in the text. Each data point represents the mean $\pm$ S.E. of values from 4-12 animals (from Gordon et al., 1988).

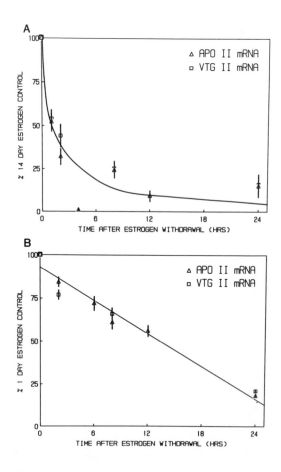

Figure 2. Decay of apoII and VTG II mRNAs. Chickens were estrogen treated, according to Fig. 1, for 14 days (A) or 24 hours (B) at which times estrogen was withdrawn by pellet removal and intramuscular injection of tamoxifen citrate (100 mg/kg). Liver RNA samples from animals sacrificed at the indicated times were assayed for apoII mRNA or VTG II mRNA as described in the text. Data are expressed as the percent ± S.E. of estrogen treated controls (from Gordon et al., 1988).

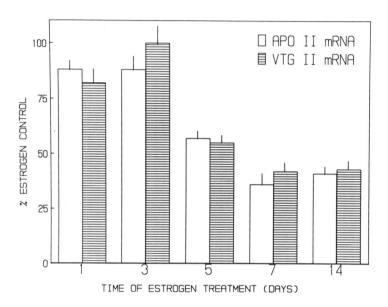

Figure 3. Induction of the apoII and VTG II mRNA destabilization activity. Eight chickens/group were treated with estrogen, according to Fig. 1, for the indicated times, after which half the chickens were sacrificed and half were withdrawn from hormone for a 3 hour period prior to being sacrificed (as described in Fig. 2). Liver RNA samples were assayed for apoII mRNA or VTG II mRNA as described in the text. Bars represent the mean percent ± S.E. of estrogen treated controls (from Gordon et al., 1988).

3 and 5 days of estrogen treatment and reaches the rapid decay mode by 7 days. This change in mRNA half life is specific to apoII and VTG II mRNAs since estrogen treatment had no effect on the turnover of apo B mRNA or on serum albumin mRNA during acute hormone withdrawal (Gordon et al., 1988). These results indicate that estrogen regulates a mechanism that selectively targets apoII and VTG

II mRNAs for degradation.

IDENTIFICATION OF SITE AND STRUCTURE SPECIFIC DEGRADATION INTERMEDIATES

We were initially interested in characterizing degradation intermediates that may accumulate during the turnover of apoII mRNA upon hormone withdrawal. Northern blot analysis of liver whole cell RNA following hormone withdrawal after 2 weeks of estrogen induction detected only full length apoII mRNA (Gordon et al., 1988). However, if degradation intermediates were very unstable *in vivo*, Northern blot analysis may not be sensitive enough to detect them. When apoII mRNA is isolated from polysomes, it is often truncated at specific sites in the 3' non-coding region (Shelness et al., 1987; MacDonald and Williams, unpublished observations). These cleavages may be a result of a polysome associated activity which may also be responsible for apoII mRNA turnover *in vivo*.

To characterize this activity we prepared 25% liver homogenates in polysome isolation buffer in the absence of detergent (25 mM Tris-HCl pH 7.5, 25 mM NaCl, 5 mM $MgCl_2$), and incubated the extract at room temperature for up to thirty minutes. If apoII mRNA is degraded in the extract by an endonucleolytic mechanism, the 3' cleavage products should be evident. Primer extension analysis from the 3' end identified a series of specific cleavage products that increased in intensity as the full length message was degraded in the extract (Fig. 4A). In Fig 4B, the presence of the corresponding 5' halves of the cleavage products mapped in Fig. 4A were verified by S1 protection mapping using a 3' end labeled S1 probe. This RNA sample was isolated from a liver extract containing 20 mM EDTA. EDTA was previously shown to enhance the *in vitro* cleavage activity (Shelness et al., 1987). When a 25% homogenate prepared with EDTA was incubated for 10 minutes at room temperature, equal amounts of full length (indicated by the arrowhead) and several truncated apoII molecules were protected by the S1 probe (lanes 5 and 6). Three fold dilution of the same extract (lanes 8 and 9) reduced the cleavage activity to the basal cleavage activity that occurs upon tissue homogenization at time 0 (not shown) such that the full length band is the major species protected. In addition, the distribution of cleavages changes when the homogenate is diluted. Upon dilution, the 3'-most cleavages, between nucleotides (nt) 587 and 637, become more prominent. This suggests that these sites are the prefered cleavage targets or that the 5'-most cleavages which predominate in the 25% homogenate, are secondary cleavages. The absence of truncated molecules in the lower portion of the gel shows that there are no cleavage sites 5' to the sites at nt 492/493. Primer extension analysis of the entire message detected only one other cluster of cleavage sites in the 5' non-coding region (not shown).

To examine the relationship between the *in vitro* nucleolytic activity and *in vivo* apoII turnover, *in vivo* RNA samples were examined by S1 protection analysis with the same 3' end labeled probe used to map the *in vitro* digested apoII mRNA. These RNA samples were isolated from livers homogenized in guanidine isothiocyanate (GTC). This is a rigorous method that should best reflect the *in vivo* state of apoII mRNA. When the poly A⁻ fractions were analyzed, a set of truncated apoII molecules similar to those mapped from liver homogenates were identified. Primer extension from the end of the message confirmed the presence of the 3' fragments of these truncated mRNAs suggesting an endonucleotytic mechanism for their generation (Binder et al., in preparation). The *in vitro* and *in vivo* cleavage sites were directly compared on primer extension and S1 mapping gels. Fig. 5A shows that four primer extension sites are present in both *in vivo* and *in vitro* digested RNA (nt 494, 520/522, 565, 600-602) while several sites are unique (nt 472 in *in vivo*, not resolved on this gel, and nt 504, 591, and 615 in *in vitro* RNA). S1 mapping (Fig. 5B) shows seven identical cleavage regions in both *in vivo* RNA and *in vitro* digestions including two (nt 593 and 611) that were not detected by primer extension of *in vivo* samples. This discrepancy may be due to either higher sensitivity of the S1 method, to instability of the 3' halves of these two cleavage products *in vivo*, or to exonuclease trimming from cleavages that occured 3' to these sites. The overall similarity of *in vivo* and *in vitro* sites suggests that the *in vitro* system will be useful for the further characterization of the mRNA destabilization activity.

The *in vitro* cleavage activity is largely restricted to the 3' non-coding region and has a distinct sequence and structure specificity. Seven out of nine cleavages mapped by S1 protection or primer extension occur at or near the trinucleotide 5'-AAU-3' within predicted looped regions on the mRNA secondary structure model (Fig. 6). Only two of the nine AAUs in the 3' non-coding region are not cleaved. The remainder of the message only contains three more AAUs in the 321 nt coding region. There are no AAUs in the 80 nt 5' non-coding region. Six of the seven cleavages mapped by primer extension occur between the A and U of the AAU. Shelness and Williams (1985) have previously shown that primer extension will accurately detect RNase T1 cleavage sites. S1 mapping accuracy will vary with DNA:RNA hybrid strength and reaction conditions so that single nucleotide resolution is probably not dependable. The S1 protection sites mapped here usually lie several nucleotides 5' to the primer extension sites, probably as a result of S1 nuclease nibbling of the probe. The seventh cleavage near an AAU (nt 636/637) could only be mapped by S1 protection because the primer extension probe hybridized too close to this site. The one primer extension site that did not occur at an AAU occured at a 5'-UA-3' dinucleotide (nt 565). The cluster of sites in the 5' non-coding region also occured within a loop at the sequence 5'-UGUAAA-3' (not shown). Therefore AU rich sequences in addition to AAU trinucleotides may also serve as nuclease targets.

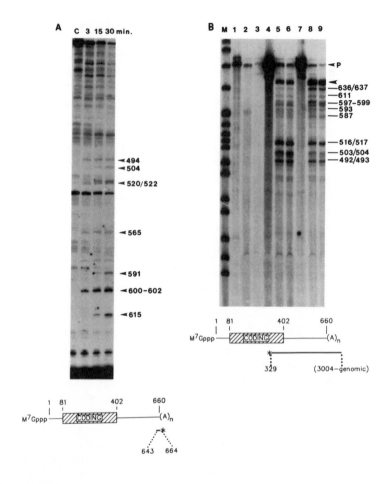

Figure 4. Analysis of *in vitro* degradation of apoII mRNA. Panel A: Equal amounts of apoII mRNA from guanidine isothiocyanate-extracted liver RNA (lane C) or RNA from 25% liver homogenates (described in the text) incubated at room temperature for 3, 15, or 30 minutes were primer extended using the 5' end-labeled primer described below the gel. The top band in all lanes represents the extension product of full length apoII mRNA (660 nucleotides) and the coordinates for shorter extension products (resulting from *in vitro* degradation) are marked along the right side of the gel. Panel B: Liver homogenates made as in Panel A, but also containing 20 mM EDTA, were incubated at room temperature for 10 minutes. The 5' digestion products were mapped by S1 nuclease protection using the 3' end-labeled probe

CONCLUSIONS

The importance of 3' non-coding regions in regulating mRNA stability has been demonstrated for a number of eukaryotic mRNAs (Brawerman, 1987). Although we have not yet demonstrated such a functional role for the apoII 3' non-coding region, the cleavage and structure data reviewed here warrants further investigation into this possibility. If the sequence and structure specific cleavages in apoII mRNA reflect the estrogen-regulated *in vivo* turnover mechanism, then several analogies to other well studied examples can be made.

The best comparison can be made with histone H4 mRNA, which is destabilized upon the completion of DNA synthesis (Sittman et al., 1983; Old and Woodland, 1984). This message is degraded by a polysome associated activity that directs the accumulation of two site specific intermediates occuring within an essential and conserved hairpin structure at the 3' end (Ross et al., 1986). Sequence specific cleavages that occur in conserved loop regions within the 3' non-coding region of apoII mRNA may also result from an activity associated with polysomes. Activation of this nuclease during polysome isolation supports this hypothesis, although direct evidence is lacking.

ApoII mRNA is the only documented example in which repeated conserved sequence/structure elements (AAU within a looped region) serve as cleavage sites for generation of degradation intermediates. In most cases, the same cleavage products induced by *in vitro* incubation of liver extracts can also be measured at lower levels *in vivo*. Most of the degradation sites were mapped by detection of both the 5' and 3' cleavage products, using S1 mapping and primer extension analysis. That one cleavage site results in two degradation products suggests an endonucleolytic mechanism, rather than a 3' to 5' exonucleolytic activity. *In vivo*, the truncated apoII molecules that result from the initial cleavage must be rapidly degraded such that they cannot accumulate to high levels. The rapid degradation may result from a 3' to 5' exonuclease. In the *in vitro* liver homogenates this rapid degradation mode may be

---

described below the gel. Lanes are: unincubated probe (1); probe annealed with or without yeast tRNA and digested with 500 u/ml of S1 nuclease (lanes 2 and 3, respectively); probe, annealed with 10 ug of liver RNA from a 25% homogenate-digestion (lanes 4-6) or a 3 fold dilution of this homogenate (lanes 7-9). For each RNA condition the annealed probe was undigested (lanes 4 and 7), treated with 500 u/ml S1 nuclease (lanes 5 and 8), or treated with 1000 u/ml S1 nuclease (lanes 6 and 9). Undigested probe (P) and the S1 resistant products reflecting full length (large arrowhead) and cleaved apoII molecules are indicated along the right side of the gel. Lane M contains pBR322/MspI markers.

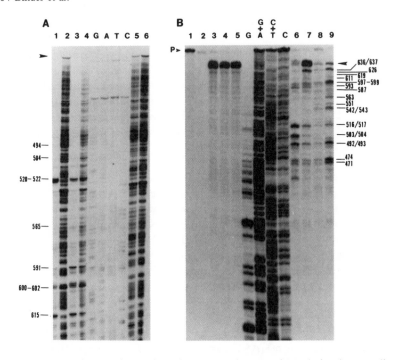

Figure 5. Comparison and mapping of *in vitro* and *in vivo* degradation intermediates. Panel A: Equal amounts of apoII mRNA were primer extended as in Fig. 4 and run next to a dideoxy sequencing ladder of an apoII M13 template. Lanes 5 and 6 represent two intact RNA samples made by the GTC extraction method. Lanes 1-4 show examples of *in vitro* (lanes 1, 3 and 4) or *in vivo* (lane 2) degradation intermediates. Lanes are: polysomal RNA (lane 1); GTC extracted RNA from a 24 hour estrogen treated and 2 hour tamoxifen withdrawn chicken (lane 2); *in vitro* digested RNA (10 minutes at room temperature) from a 25% homogenate with EDTA (lane 3) or a three fold dilution of this homogenate (lane 4). Panel B: The same samples that were analyzed in Panel A were S1 mapped, as described in Fig. 4. The S1 protected fragments were run next to a Maxam and Gilbert sequencing ladder of the S1 probe. Lanes are: probe annealed with yeast tRNA, undigested (lane 1), or digested with S1 nuclease (lane 2); probe annealed with RNA from the poly A$^+$ fraction (lane 4) and poly A$^-$ fraction (lane 8) from 24 hour estrogen treated chickens; probe annealed with RNA from the poly A$^+$ fraction (lane 5) and poly A$^-$ fraction (lane 9) from 24 hour estrogen treated and 2 hour tamoxifen withdrawn chickens; probe annealed with liver RNA prepared from a 2 week estrogen induced chicken by GTC extraction (lane 3) or following *in vitro* digestion, as described in Panel A, of a 25% homogenate (lane 6) or a 3-fold dilution of the homogenate (lane 7). For each RNA sample the annealed probe was digested with 2000 u/ml S1 nuclease.

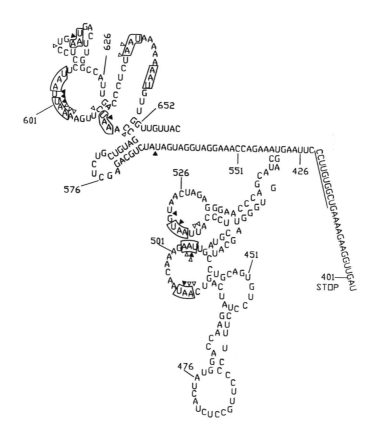

Figure 6. *In vitro* digestion cleavage products occur at or near the trinucleotide 5'-AAU-3' within loop regions on a secondary structure model of the apoII 3' non-coding region. 5' cleavage products mapped by S1 protection (▽) and 3' cleavage products mapped by primer extension (▼ ; taken from Fig. 4) are shown on the secondary structure model. AAUs are boxed. The underlined region, spanning nucleotides 414-423, is base paired to the coding region in a larger model from which the region shown here was taken (Hwang, et al., 1988).

slowed or inactivated, facilitating accumulation of the endonuclease cleavage products. The initial endonucleolytic cleavage may be the rate limiting step which is regulated *in vivo* when apoII turnover switches from the slow to rapid decay mode. Lack of qualitative differences in degradation intermediates from the two turnover modes supports this hypothesis (data not shown).

The secondary structure context in which the AAUs occur may be very important for directing regulated cleavages to these sites. Putative stabilizing or nucleolytic regulatory proteins may recognize this particular structure/sequence motif. Apo B mRNA turnover is an example where the presence of AAU trinucleotides are not sufficient to confer an estrogen-induced stability change. Apo B is also transcriptionally induced by estrogen treatment in chicken liver. Although the apoB mRNA 3' non-coding region is AAU rich, its 3 hour half life is not altered by the length of prior estrogen treatment (Kirchgessner et al., 1987; Gordon et al., 1988). Perhaps AAUs can confer general instability to mRNAs such as apo B. The unique 1° and 2° structure of apoII mRNA may act as a recognition signal for regulatory proteins that can further modulate accesibility to destabilizing factors. We have preliminary evidence showing specific protein binding to the 3' non-coding region of apoII (Ratnasabapathy and Williams, in preparation). This is analogous to ferritin mRNA where a conserved stem loop structure in the 5' non-coding region is required for translational regulation and specific protein binding (Aziz and Munro, 1987; Hentze et al., 1987; Leibold and Munro, 1988; Rouault et al., 1988).

There are several other examples in which 3' non-coding AU sequences may modulate message stability. An AU-rich repeating sequence (AUUUA) has been identified in the 3' non-coding regions of a number of unstable messages, such as GM-CSF, c-fos, c-myc, cachectin (TNF), and interferon (Caput et al., 1986; Shaw and Kamen, 1986). Transfection studies have shown that several of these 3' non-coding regions can confer instability to either the homologous message or to heterologous stable messages (Shaw and Kamen, 1986; Schuler and Cole, 1988; Wilson and Treisman, 1988). In contrast, the stable message for globin contains 3' non-coding region AU sequences which serve as nuclease targets but do not confer a short half life to the message (Albrecht et al., 1984). These data suggest that AU-rich sequences may be modulating message stability in a sequence context fashion which may also involve specific secondary structure and protein interaction requirements. Functional studies of the apoII 3' non-coding region will serve to distinguish mRNA elements necessary for turnover from those which regulate the rate of turnover.

ACKNOWLEDGEMENT

This work was supported in part by Grant DK 18171 from the National Institutes of Health. R.B. was supported by N.I.H. Predoctoral Award (National Research Service Award) GM 08065. D.A.G. was supported by N.I.H. Predoctoral Award (National Research Service Award) GM 07518.

REFERENCES

Albrecht G, Krowczynska A, Brawerman G (1984). Configuration of ß-globin messenger RNA in rabbit reticulocytes. J Mol Biol 178:881-896.

Aziz N, Munro, HN (1987). Iron regualtes ferritin mRNA translation through a segment of its 5' untranslated region. Proc Natl Acad Sci USA 84:8478-8482.

Berlin CM, Schimke RT (1965). Influence of turnover rates on the responses of enzymes to cortisone. Mol Pharmacol 1:149-156.

Binder R, Hwang SL, Williams DL (1988). In preparation.

Brawerman G (1987). Determinants of messenger RNA stability. Cell 48:5-6.

Brock ML, Shapiro DJ (1983a). Estrogen regulates the absolute rate of transcription of the Xenopus laevis vitellogenin genes. J Biol Chem 258:5449-5455.

Brock ML, Shapiro DJ (1983b). Estrogen stabilizes vitellogenin mRNA against cytoplasmic degradation. Cell 34:207-214.

Caput D, Beutler B, Hartog K, Thayer R, Brown-Shimer S, Cerami A (1986). Identification of a common nucleotide sequence in the 3' untranslated region of mRNA molecules specifying inflammatory mediators. Proc Natl Acad Sci USA 83:1670-1674.

Gordon DA, Shelness GS, Nicosia M, Williams DL (1988). Estrogen-induced destabilization of yolk precursor protein mRNAs in avian liver. J Biol Chem 263:2625-2631.

Hentze MW, Wright Caughman S, Rouault TA, Barriocanal JG, Dancis A, Harford JB, Klausner RD (1987). Identification of the iron-responsive element for the translational regulation of human ferritin mRNA. Science 238:1570-1573.

Hwang SL, Eisenberg M, Binder R, Shelness GS, Williams DL (1988). Submitted for publication.

Hynes NE, Groner B, Sippel AE, Jeep S, Wurtz T, Nguyen-Huu MC, Giesecke K, Schütz G (1979). Control of cellular content of chicken egg white protein specific RNA during estrogen administration and withdrawal. Biochem 18:616-624.

Kirchgessner TG, Heinzmann C, Svenson KL, Gordon DA, Nicosia M, Lebherz HG, Lusis AJ, Williams DL (1987). Regulation of chicken apolipoprotein B: cloning, tissue distribution, and estrogen induction of mRNA. Gene 59:241-251.

Leibold EA, Munro HN (1988). Cytoplasmic protein binds in vitro to a highly conserved sequence in the 5' untranslated region of ferritin heavy-and light-subunit mRNAs. Proc Natl Acad Sci USA 85:2171-2175.

Old RW, Woodland HR (1984). Histone genes: not so simple after all. Cell 38:624-626.

Palmiter RD, Carey NH (1974). Rapid inactivation of ovalbumin messenger ribonucleic acid after acute withdrawal of estrogen. Proc Nat Acad Sci USA 71:2357-2361.

Ratnasabapathy R, Williams DL (1988). In preparation.

Ross J, Peltz SW, Kobs G, Brewer G (1986). Histone mRNA degradation *in vivo*: the first detectable step occurs at or near the 3' terminus. Mol Cell Biol 6:4362-4371.

Rouault TA, Hentze MW, Wright Caughman S, Harford JB, Klausner RD (1988). Binding of a cytosolic protein to the iron-responsive element of human ferritin messenger RNA. Science 241:1207-1210.

Schuler GD, Cole MD (1988). GM-CSF and oncogene mRNA stabilities are independently regulated in trans in a mouse monocytic tumor. Cell 55:1115-1122.

Shapiro DJ, Baker HJ (1977). Purification and characterization of Xenopus laevis vitellogenin messenger RNA. J Biol Chem 252:5244-5250.

Shapiro DJ, Blume JE, Nielsen DA (1987). Regulation of messenger RNA stability in eukaryotic cells. BioEssays 6:221-226.

Shaw G, Kamen R (1986). A conserved AU sequence from the 3' untranslated region of GM-CSF mRNA mediates selective mRNA degradation. Cell 46:659-667.

Shelness GS, Binder R, Hwang SL, MacDonald C, Gordon DA, Williams DL (1987). Sequence and structural elements associated with the degradation of apolipoprotein II messenger RNA. In Inouye M, Dudock BS (eds): "Molecular Biology of RNA: New Perspectives," San Diego: Academic Press Inc., pp 381-399.

Shelness GS, Williams DL (1985). Secondary structure analysis of apolipoprotein II mRNA using enzymatic probes and reverse transcriptase. J Biol Chem 260:8637-8646.

Sittman DB, Graves RA, Marzluff WF (1983). Histone mRNA concentrations are regulated at the level of transcription and mRNA degradation. Proc Natl Acad Sci USA 80:1849-1853.

Watson G, Davey RA, Labarca C, Paigan K (1981). Genetic determination of kinetic parameters in β-glucuronidase induction by androgen. J Biol Chem 256:3005-3011.

Wilson T, Treisman R (1988). Removal of poly(A) and consequent degradation of c-fos mRNA facilitated by 3' AU-rich sequences. Nature 336:396-399.

Wiskocil R, Bensky P, Dower W, Goldberger RF, Gordon JI, Deeley RG (1980). Coordinate regulation of two estrogen-dependent genes in avian liver. Proc Natl Acad Sci USA 77:4474-4478.

V. STEROID HORMONE RECEPTORS IN CANCER

Molecular Endocrinology and Steroid
Hormone Action, pages 243–278
© 1990 Alan R. Liss, Inc.

MECHANISMS OF HORMONE AND CYTOTOXIC DRUG INTERACTIONS IN THE DEVELOPMENT AND TREATMENT OF BREAST CANCER

Robert Clarke, Marc E. Lippman and Robert B. Dickson

Lombardi Cancer Research Center (S128), Georgetown University Medical Center, 3800 Reservoir Road NW, Washington, DC 20007, USA

INTRODUCTION

Breast cancer is the most common form of cancer among women in western society, the disease being predominately associated with middle and old age. Approximately 1 in 11 of all women in the USA will develop breast cancer, male breast cancer has an incidence 1% of that in women (Nirmul et al., 1982). Breast carcinomas arise largely from the basal cells of the ductal epithelium, spread radially by infiltrating through tissue spaces and can invade both lymphatic and blood vessels. There is a high incidence of distant metastases, particularly in soft tissues, lungs, liver, bones and adrenals. In the management of breast cancer the choice of treatment depends largely upon the stage of the disease at diagnosis. Where there is no evidence of metastasis, treatment has traditionally been confined to local modalities such as surgery and radiotherapy. However, the curative potential of these regimens is restricted by the disseminated nature of the disease. At diagnosis over 60% of patients may have either occult or clinical metastatic disease and, therefore, require some form of systemic treatment such as endocrine manipulation or chemotherapy (Henderson & Canellos, 1980). Furthermore, whilst most metastatic tumors initially respond to systemic treatment, the overwhelming majority progress to a phenotype characterized by resistance to both cytotoxic drugs and hormonal therapies.

The development of new therapies for breast cancer may require a more detailed understanding of the mechanisms involved in cellular transformation, malignant progression and tumor growth. Many of the factors responsible for the control of normal breast growth and development are inextricably linked with the control of malignant breast tissue. We shall describe the role of hormones and growth factors in the growth of both normal and malignant breast tissue, and examine the interactions between hormonal and chemotherapeutic treatments in breast cancer.

GROWTH AND DEVELOPMENT OF NORMAL BREAST TISSUE

Role of hormones

The precise hormonal interactions ultimately responsible for the proliferation and differentiation of normal breast tissue remain unclear. Whilst a number of hormones have been implicated in mammary development, those most closely associated include pituitary factors, insulin and the steroids. Various stages of differentiation and development often require complex combinations of these hormones.

Estrogen, progesterone and hydrocortisone are amongst the most influential steroids involved in mediating the growth and differentiation of normal mammary tissue. Estrogens increase nipple differentiation and induce proliferation of the surrounding mesenchyme in the developing mouse fetus (Raynaud, 1955). The ductal and lobulo-alveolar phases of development are induced by estradiol in ovariectomized but not in ovariectomized\ hypophysectomized mice (Lyons, 1958; Lieberman et al., 1978) indicating the necessity for additional pituitary factors. In the adult mammary gland DNA synthesis in epithelial cells reaches its maximum during the estrogenic phase of the estrous cycle (Sutton & Suhrbrier, 1974). Estrogen and progesterone increase alveolar formation and ductal branching (Bronson et al., 1975; Murr et al., 1974). *In vitro* studies of mammary development indicate that, for full lobulo-alveolar development, tissue must first be primed by exposure to both estrogen and progesterone (Warner, 1978). The mechanism by which estrogen influences the growth and differentiation of normal breast tissue remains to be elucidated. However, it may be either a direct effect or the result of an indirect interaction mediated through stromal tissues. For example, estrogen only induces epithelial proliferation *in vitro* in the presence of mammary stromal cells (McGrath, 1983; Haslem & Levely, 1985). A direct effect would be mediated by interaction with the specific cellular receptors for estrogen which have been observed in normal human breast tissue (Petersen et al., 1987; Balakrishnan et al., 1987).

Insulin, hydrocortisone and prolactin increase the number of functionally differentiated cells during alveolar differentiation. Although all of these hormones are required for the maximal response, either insulin or epidermal growth factor (EGF) alone can stimulate cell differentiation and division in the proliferative stage (Turkington, 1971). Furthermore, there is a quantitative relationship between alveolar development and the mitogenic responses to insulin, prolactin, estrogen and progesterone in mouse mammary epithelial cells *in vitro* (Dilley, 1971). Following development, insulin alone is sufficient to maintain mouse mammary glands *in vitro*, ducts remain viable (Mehta et al., 1975) but full alveolar regression occurs (Wood et al., 1975). Induction of milk protein synthesis requires insulin, prolactin and hydrocortisone (Terry et al., 1975; Topper & Freeman, 1980).

Role of growth factors.

Both EGF and transforming growth factor alpha (TGF-alpha), a naturally occurring homologue of EGF, have been isolated from human milk (Zwiebel et al., 1986). The submaxillary glands are the main source of circulating EGF in mice. Sialoadenectomized lactating mice have smaller mammary tissue and reduced milk production than sham-operated mice indicating a possible role for EGF in the development of mammary tissue during pregnancy (Oka et al., 1987). During the proliferative stages of alveolar differentiation EGF can stimulate cell differentiation and division (Turkington, 1971). Both EGF and TGF-alpha stimulate lobulo-alveolar development *in vitro*. However, TGF-alpha appears to be more potent than EGF. Localized mammary growth in Balb\c mice is stimulated by EGF but a simultaneous exposure to both estrogen and progesterone is required. In contrast, TGF-alpha but not EGF is capable of stimulating growth in C3H/HeN mice without steroid supplementation, although, maximal growth induction by TGF-alpha is observed in the presence of steroids (Vondehaar, 1988). Human mammary epithelial cells require EGF for continued growth *in vitro*, especially at low density (Stampfer et al., 1980).

Insulin-like growth factors (IGFs) may be involved in normal breast development. Rosenfeld et al. (1983) observed a correlation between changes in breast size and circulating levels of IGF-I. Secretion of both IGF-I and IGF-II increases during lactation in bovine mammary tissue (Dehoff et al., 1988). Receptors for IGF-I have been detected in normal human breast tissue (Pekonen et al., 1988). High levels of insulin are required for the maintenance of normal rat mammary epithelial cells in culture. However, IGF-I can replace insulin at much lower concentrations, perhaps indicating action through an IGF-I-mediated mechanism (Deeks et al., 1988). IGF-II mRNA is expressed by the stromal component of some breast tumors (Cullen et al., 1988).

GROWTH AND DEVELOPMENT OF MALIGNANT BREAST TISSUE

Role of hormones.

Many of the hormones associated with the control of differentiation and development in normal breast tissue also influence the growth of neoplastic breast. Exceptions include both prolactin and progesterone. Whilst all rodent mammary tumors appear, at least initially, dependent upon prolactin (Briand, 1983), no significant role has yet been established for prolactin in human breast cancer (recently reviewed by L'Hermite & L'Hermite-Baleriaux, 1988).

Progestins are generally considered to be differentiating agents and their role in the genesis and control of human breast cancer remains unclear. However, pharmacological concentrations of medroxyprogesterone acetate induce remissions in approximately 30% of patients, a response rate similar to that observed following treatment with the antiestrogen tamoxifen. Perturbations in estrogen receptor (ER) expression, both reduced steroid synthesis and testosterone breakdown, and an increase in the conversion of estradiol to estrone, have been implicated in the mechanism of action of pharmacological concentrations of progestins (recently reviewed by Harmsen & Porsius, 1988). The ability to predict response to antiestrogen therapy is increased when progesterone receptor (PGR) is included with ER determinations (Clark & McGuire, 1983; King et al., 1982). However, this probably reflects the ability of PGR expression to indicate the presence of a functional ER system.

Estrogen is the hormone most closely associated with the control of breast tumor growth. For example, estrogen plays a major role in the ductal phases of growth in normal breast and it is from the basal cells of the ductal epithelium that most breast cancers are thought to arise. However, the precise role of estrogens remains unclear, although, a potential relationship between ovarian function and breast cancer has been postulated for some time. The Italian epidemiologist Ramazzini (1633-1714) observed a higher incidence of breast cancer in nuns compared with married women. Cooper (1835) reported a correlation between cyclic variations in the size of breast tumors and the phase of the menstrual cycle. Whilst Schinzinger (1889) suggested that removal of the ovaries might cause a slowing of the growth of breast tumors, it was Beatson (1896) who first reported remissions in premenopausal breast cancer patients following bilateral ovariectomy. A potential role for estrogens in the induction of mammary cancers was observed by Lacassagne (1932) when he chronically administered an ovarian extract containing estrone benzoate to castrated male mice and subsequently induced mammary tumor formation. More recent evidence indicating a central role for estrogen in the control of breast cancer proliferation includes the inhibition by antiestrogens of tumors expressing cellular receptors for both estrogen and progesterone (Clark & McGuire, 1983; King et al., 1982). Human breast cancer cells expressing these receptors are both inhibited by pharmacological concentrations of antiestrogens and stimulated by physiological concentrations of estrogens *in vitro* (Lippman et al., 1976; Aitken & Lippman, 1982). Furthermore, these cell lines require estrogen for tumor formation in athymic nude mice (Soule & McGrath, 1980; Shafie & Grantham, 1981; Siebert et al., 1983;).

Estrogens could function as carcinogens, co-carcinogens and\or tumor promoters. A number of observations, including their ability to influence the rate of proliferation, atrophy or differentiation of stem or intermediate cells in normal breast, may reflect the function of tumor promoters rather than mutagens (Thomas, 1984). The bioavailability of serum estradiol does not account for the varying rates of breast cancer observed among normal Caucasian, Chinese, Hawaiian and Japanese women living in Hawaii (Goodman et al., 1988). The incidence of rat mammary tumors induced by 7,12-dimethylbenz(a)anthracene is significantly reduced if estrogens are removed before or shortly after administration of the drug (Dao, 1972). Estrogen increases the rate of tumor formation in Wistar\Furth rats infected with the adenovirus type 9 (Ankerst & Jonsson, 1989). The reduced risk of breast cancer in women who smoked heavily as teenagers has been attributed to the smoking-induced reduction in estrogen production or altered estrogen metabolism (Michnovicz et al., 1986). The ability of estrogens to stimulate the rate of proliferation of breast cancer cells *in vitro* and of ovariectomy and antiestrogens to induce remissions in some patients also strongly support a role as tumor promoter (Thomas, 1984). However, duration of exposure to estrogens has been associated with endometrial (Jick et al., 1980) and liver cancers (Vana et al., 1979) and vaginal adenocarcinoma (Herbst et al., 1971). In women, breast cancer occurs after puberty, although, a rare secretory carcinoma of the breast has been reported in both children and adolescent girls (Akhtar et al., 1983). The risk of breast cancer is reduced following premenopausal ovariectomy without estrogen replacement and there is an inverse relationship between the degree of protection and age at ovariectomy (Thomas, 1984). Moreover, in the case of primary ovarian failure the incidence of breast cancer approaches that observed in men. These observations suggest that estrogen may also function as a carcinogen or co-carcinogen.

Paradoxically, pharmacological concentrations of estrogens such as diethylstilbestrol can inhibit some estrogen-dependent breast tumors. Until the advent of tamoxifen, administration of diethylstilbestrol was the additive endocrine manipulation of choice. Response rates of up to 38% have been achieved in unselected breast cancer patients (Henderson, 1987). These response rates are approximately equivalent to those obtained following administration of antiestrogens. In contrast to antiestrogen therapy, diethylstilbestrol induces considerable host toxicity. The mechanism of action of high concentrations of steroids is unclear but non-specific effects mediated by changes in plasma membrane fluidity may be involved (Clarke et al., 1987).

Insulin plays an important role in the growth and development of both normal and malignant breast. In human breast cancer cells *in vitro*, insulin stimulates fatty acid (Monaco & Lippman, 1977) DNA, RNA and protein synthesis and cell division (Pilkis & Park, 1974; Osborne et al., 1976). The majority of 7,12- dimethylbenz(a)anthracene-induced mammary tumors are insulin dependent and this dependency appears to be mediated through a process related to DNA synthesis (Heuson & Legros, 1970). Mammary tumor formation can be prevented by alloxan-induced diabetes 3-4 weeks after treatment with 7,12-dimethylbenz(a)anthracene (Heuson & Legros, 1972) and is restored by the administration of insulin (Heuson & Legros, 1970). Approximately 60% of 7,12-dimethylbenz(a)anthracene-induced rat mammary tumors regress when the animals are made diabetic following treatment with streptozotocin. Regression is reversed upon administration of insulin (Cohen & Hilf, 1974). MCF-7 human breast cancer cells require insulin for tumor formation and growth *in vivo* in the athymic nude mouse. Streptozotocin-induced diabetes both prevents tumor formation and induces MCF-7 tumor regression. Under these conditions tumor growth can be restored by the administration of insulin but not estrogen (Shafie & Grantham, 1981). Mouse mammary aplastic carcinomas, after several passages between alloxan-diabetic rats, become conditioned to hyperinsulinemia by secreting substances immunologically cross-reactive with insulin (Vuk-Pavlovic et al., 1982).

LH-RH analogs have been reported to directly inhibit MCF-7 cells *in vitro* (Fockens et al., 1986). When administered in combination, the analogs Pro 9-LH-RH ethylamide and D Ser(t BU)6 Aza Gly 10-LH-RH (ICI 118630) inhibit the estrogen-induced stimulation of proliferation in CG-5 cells, a subline of the MCF-7 human breast cancer cell line (Scambia et al., 1988). However, the major mechanism of action of this class of compounds *in vivo* is more likely to be related to their ability to inhibit gonadotrophin release and the consequent reduction in secreted gonadal steroids (Fur & Nicholson, 1982).

Both somatostatin analogs and the parent hormone inhibit the growth of some mammary tumors *in vivo* (Vuk-Pavlovik et al., 1982; Defeudis & Moreau, 1986) however, the precise mechanism of action is unknown. Somatostatin analogs inhibit breast cancer cells *in vitro* (Setyono-Han et al., 1987; Cremin et al. in press). Furthermore, these analogs reduce plasma concentrations of IGF-I (Lamberts et al., 1985) and EGF (Ghirlanda et al., 1983). Thus, inhibition of tumor growth may be the result of a combination of indirect endocrine and direct antitumor effects.

Role of growth factors.

The proliferation of both human (Barnes & Sato, 1978; Osborne et al., 1980; Imai et al., 1982) and rodent breast cancer cells (Turkington, 1969) is stimulated by EGF\TGF-alpha. Sialoadenectomy reduces the incidence of mammary tumors in mice which have a high incidence of spontaneous mammary tumor formation. EGF replacement therapy increases the level of successful transplantation of these tumors (Oka et al., 1987; Tsutsumi et al., 1987). EGF can increase the ability of estrogen dependent MCF-7 human breast cancer cells to form tumors in ovariectomized nude mice (Dickson et al., 1986). Approximately 70% of human breast tumors express TGF-alpha mRNA (Bates et al., 1988) and a number of breast cancer cell lines have been reported to produce both TGF-alpha mRNA and protein (Bates et al., 1988; Peres et al., 1987; Perroteau et al., 1986; Clarke et al., in press). TGF-alpha-like material has been isolated from the urine of breast cancer patients (Sherwin et al., 1983; Kimball et al., 1984; Stromberg et al., 1987). The urines of both normal and tumor-bearing mice also contain TGF-alpha-like material (Twardzik et al., 1985).

Immunocytochemical analysis has revealed the presence of insulin-like material in some human breast tumors (Spring-Mills et al., 1984). IGF-I stimulates the proliferation of some breast cancer cells (Furlanetto & DiCarlo, 1984; Huff et al., 1986) and can induce the formation of hormone independent MCF-7 tumors in ovariectomized athymic mice. However, it is less potent than EGF for induction of tumorigenesis (Dickson et al., 1987). The growth of MDA-MB-231 cells both *in vivo* and *in vitro* is inhibited by an antibody which blocks ligand binding to the IGF-I receptor (Rohlik et al., 1987; Arteaga et al., 1988). This antibody also inhibits the growth of MCF-7 cells *in vitro* (Rohlik et al., 1987). IGF-II mRNA is expressed in T47D cells and stimulates the growth of MCF-7 and T47D breast cancer cells *in vitro* (Cullen et al., 1988). Since the majority of IGF-II mRNA is expressed in non-malignant areas of some tumors (Cullen et al., 1988), IGF-II may be both an autocrine and a paracrine growth factor in breast cancer.

COMMA-D breast epithelial cells respond mitogenically to basic fibroblast growth factor (Riss & Sirbasku, 1987). Some human breast cancer cell lines produce a 60 kilodalton basic fibroblast growth factor-like activity (Swain et al., 1986). Whilst the precise nature of this material remains unclear, it may contribute, in conjunction with other mitogenic factors, to the control of proliferation.

Transforming growth factor ß (TGF-ß) is secreted by a number of breast cancer cell lines (Knabbe et al., 1987; Arteaga et al., 1988). However, in contrast to the factors mentioned above, TGF-ß inhibits the proliferation of many breast cancer cells. Thus, an additional level of control of proliferation may result from altered secretion of potential inhibitory factors.

Hormone\growth factor interactions.

In breast cancer cells, the level of secretion of a number of growth factors is altered by estrogen. For example, TGF-alpha levels are increased two fold or greater following treatment of MCF-7 cells with physiological concentrations of 17ß-estradiol (Dickson et al., 1986; Bates et al., 1988). Estrogen increases secretion of an IGF-I-like material in MCF-7 cells (Huff et al., 1988) and IGF-II mRNA in T47D cells (Yee et al., 1988). A small induction in platelet derived growth factor (PDGF) mRNA production and a larger induction of PDGF biological activity has also been observed (Bronzert et al., 1988). Secretion of an inhibitory growth factor TGF-beta is reduced by estrogen but increased by antiestrogens (Knabbe et al., 1987).

The production of EGF in mouse submaxillary glands is modulated by a number of hormones including both progestins and androgens (Cohen, 1983). EGF in combination with insulin, prolactin, aldosterone and hydrocortisone is required for the further development of mouse mammary glands *in vitro* following complete alveolar regression (Tonelli et al., 1980).

Insulin and EGF are at least additive in stimulating the growth of MCF-7 cells (Barnes & Sato, 1978; Huff et al., 1988) COMMA-D mouse mammary epithelial cells (Riss & Sirbasku, 1987) and normal rat mammary epithelial cells (Ethier et al., 1987). Both exogenous TGF-alpha and EGF can increase production of an IGF-I-related material in MCF-7 cells (Huff et al., 1988) and IGF-I synergizes with estrogen in stimulating MCF-7 tumor formation (Huff et al., 1988). Thus, either insulin and/or IGF-related proteins may act together with TGF-alpha in modulating the proliferation of breast tumor cells.

Estrogen and insulin interact in a variety of mammary tumor cells. Insulin increases estrogen binding to R3230AC rat mammary adenocarcinoma cells *in vitro* (Shafie et al., 1977). Removal of administered insulin from diabetic rats decreases estrogen binding to these cells (Hilf et al., 1978). Insulin-dependent and -independent tumors in diabetic rats express increased ER levels in response to treatment with physiological concentrations of insulin (Hilf et al., 1978). Estrogen and insulin are synergistic in stimulating the proliferation of MCF-7 cells grown in media supplemented with serum treated with dithiothreitol and iodoacetamide to destroy most growth factor activity (van der Berg et al., 1987). In contrast, pharmacological or greater concentrations of insulin reduce both ER expression in MCF-7 cells *in vitro* (Moore, 1980) and the effects of both estrogens and antiestrogens (Butler et al., 1981).

Both EGF (Murphy et al., 1988) and EGF-R expression (Murphy et al., 1986; Sarup et al., 1988) are increased in T47D cells and EGF-R expression in MCF-7 and ZR-75-1 cells (Leake et al., 1988) following treatment with progestins. EGF reduces progesterone binding and the inhibitory effects of the synthetic progestin R5020 (Sarup et al., 1988). Prolactin can induce a 10-fold increase in EGF-R expression in MCF-7 and ZR-75-1 cells (Leake et al., 1988). Estrogen increases EGF-R protein (Mukku & Stancel, 1985) and mRNA levels (Lingham et al., 1988) in rat uterine tissue.

EGF partially reverses the growth inhibitory effects of the antiestrogens 4-hydroxytamoxifen, tamoxifen, 4-hydroxyclomiphene and LY117018 (Koga & Sutherland, 1987). Furthermore, both insulin- and EGF-stimulated cell proliferation in MCF-7 cells growing in the absence of estrogen is inhibited by antiestrogens (Vignon et al., 1987).

Hormone\growth factor receptors.

The ability of a cell to respond to a growth factor or hormone is generally the result of an interaction between the relevant ligand and a specific, high affinity, low capacity receptor located predominantly either in the plasma membrane e.g. insulin or EGF receptors or intracellularly e.g. steroid receptors. Consequently, the presence of receptors may indicate a specific requirement of the cell for a hormone or growth factor. The ligand may be obtained from the same cell, from surrounding tissue, or from distant secreting tissues.

ER and PGR are the most important hormone receptors

associated with breast cancer. Approximately 30% of all breast tumors respond to the antiestrogen tamoxifen (Rose et al., 1982; Bradbeer & Kyngdom, 1983) with a greater proportion of responding tumors being ER positive (Carter, 1981; Clarysse, 1985). Combined measurement of both ER and PGR increases the ability to predict initial response to endocrine therapy to approximately 60% or greater (King et al., 1982; Stewart et al., 1982; Clark & McGuire, 1983). Approximately 10% of ER negative tumors respond to endocrine manipulation, perhaps reflecting the heterogeneity of the tumors and\or the sensitivity of the ER assays. Henry et al. (1988) compared ^{3}H-estradiol binding capacity with ER mRNA expression and reported that a high proportion of ER negative tumors (<5 fmol\mg ER protein) express detectable levels of ER mRNA.

High body weight is associated with increased production of estrogens but not ER levels in patients with advanced breast cancer. However, PGR expression correlates with high body weight, particularly in patients with ER positive tumors (Williams et al., 1988).

Attempts to correlate ER content with either tumor size (Heuson & Benraad, 1983) or the time of onset or incidence of distant metastases (Campbell et al., 1981; Kamby et al., 1988) have been largely unsuccessful. However, low ER levels and a high degree of anaplasia are frequently associated with visceral metastases, whilst tumors expressing high levels of ER often metastasize to bone (Kamby et al., 1988). Histological grade III tumors are more likely to be ER negative compared with either grade I or grade II tumors (Blanco et al., 1984; Singh et al., 1988; Henry et al., 1988). ER expression is a poor indicator of response to cytotoxic drugs. Kinne et al. (1981) and Stewart et al. (1982) reported no difference in response to chemotherapy between patients with ER positive and ER negative tumors. A number of investigators have observed a close correlation between ER expression and tumor growth rate. Tumors with low ER content proliferate more rapidly than tumors expressing high levels of ER (Jonat & Maat, 1978; Antoniades & Spector, 1982; Meyers et al., 1977).

The value of ER expression as a prognostic indicator is equivocal (Patterson et al., 1982), although, ER negative tumors tend to indicate a worse prognosis (Godolphin et al., 1981; Aamdal et al., 1984; Skoog et al., 1987). The most reliable prognostic indicators in breast cancer remain clinical stage and pathological lymph node status (Aamdal et al., 1984; Clark & McGuire, 1988).

There appears to be an inverse relationship between ER and EGF-R expression in malignant breast tissue. Primary breast tumors (Sainsbury et al., 1985; Perez et al., 1984;

Cattoretti et al., 1988; Wrba et al., 1988; Pekonen et al., 1988) and human breast cancer cell lines (Davidson et al., 1987) which have either low ER content or lost the ability to express ER possess high levels of EGF-R. Furthermore, metastases frequently express more EGF-R than primary tumors (Sainsbury et al., 1985). EGF-R expression is directly associated with an increased risk of early recurrence and death in all breast cancer patients (Sainsbury et al., 1988) and with failure of primary endocrine therapy in postmenopausal patients (Nicholson et al., 1988). A high level of EGF-R has also been reported to be an indicator of poor prognosis in bladder (Neal et al., 1985) colon (Bradley et al., 1986) and non-small cell lung cancers (Veale et al., 1987). Whilst overexpression of EGF-R is a more reliable prognostic indicator than ER status (Sainsbury et al., 1985; Sainsbury et al., 1987) a recent study failed to observe any significant correlation between EGF-R expression and growth fraction, grading, tumor diameter and lymph node status in 88 primary breast carcinomas (Wrba et al., 1988).

In contrast to the relationship between EGF-R and ER, expression of IGF-I receptor correlates directly to ER and PGR expression (Pekonen et al., 1988; Peyrat et al., 1988).

Breast cancer cells also express a variety of receptors to growth factors and hormones in addition to EGF-R, ER and PGR. For example, functional receptors for glucocorticoids, dihydrotestosterone (Horwitz et al., 1975) prolactin (Shafie & Brooks, 1977) insulin (Osborne et al., 1978) IGF-I and IGF-II (Furlanetto & DiCarlo, 1984; Myal et al., 1984; De Leon et al., 1988) and glucagon (Cremin et al., 1987) have been reported on some breast cancer cell lines. T47D cells express receptors for vasoactive intestinal polypeptide (Gespach et al., 1988) and some human breast carcinomas express gonadotropin-releasing hormone receptors (Eidne et al., 1985).

THE AUTOCRINE HYPOTHESIS: A UNIFYING CONCEPT FOR THE CONTROL OF BREAST TUMOR GROWTH

In addition to possessing receptors for a number of mitogenic growth factors some breast cancer cells also secrete these growth factors. Hence, an autocrine stimulatory mechanism may control breast tumor proliferation. For example, hormone independent cells have been reported to constitutively secrete TGF-alpha (Bates et al., 1988; Peres et al., 1987; Perroteau et al., 1986) and to express EGF-R (Davidson et al., 1987; Fitzpatrick et al., 1984). Whilst considerable attention has focused on TGF-alpha as the main mediator of these effects, it's precise role in breast cancer growth remains unclear (Clarke et al., in press). The greater sensitivity of normal breast to TGF-alpha, as outlined previously, may indicate that TGF-alpha rather than EGF is the "natural" ligand for the EGF receptor (EGF-R) in developing breast tissue. This hypothesis gains further support from the observation that, whilst a number of human breast cancer cell lines have been reported to produce TGF-alpha (Bates et al., 1988; Peres et al., 1987; Perroteau et al., 1986), only one breast cancer cell line to date has been reported to produce detectable levels of EGF (Murphy et al., 1988). Thus, in some breast tissue TGF-alpha may function as an onco-fetal protein or as a marker for malignancy.

Secretion of TGF-alpha is regulated by estrogen in ovarian dependent human breast cancer cell lines (Bates et al., 1988). Thus, TGF-alpha may be involved in estrogen regulated growth. Hormone dependent MCF-7 cells transfected with a TGF-alpha expression vector constitutively secrete levels of biologically active TGF-alpha equivalent to hormone independent MDA-MB-231 cells. However, these cells fail to form tumors in ovariectomized athymic mice without estrogen supplementation (Clarke, et al., in press). Furthermore, a subcutaneous administration of 1 microgram\day EGF in ovariectomized nude mice is not sufficient to fully support the growth of MCF-7 tumors (Dickson et al., 1987). Thus, TGF-alpha expression alone is not sufficient to fully mediate the effects of estrogen *in vivo*.

The ability of estrogens and progestins to increase EGF-R expression in estrogen-responsive tissue (Mukku & Stancel, 1985; Lingham et al., 1988; Leake et al., 1988) may indicate an additional requirement for elevated numbers of EGF cell surface receptors for mediation of the full effects of TGF-alpha secretion. Alternatively, TGF-alpha may function in combination with other estrogen regulated mitogenic growth factors. For example, estrogen can stimulate the secretion of IGF-like materials in some breast cancer cells (Huff et al., 1988; Yee et al., 1988). Since insulin and TGF-alpha are at least additive in stimulating the proliferation of breast cancer cells (Barnes & Sato, 1978; Huff et al., 1988; Riss & Sirbasku, 1987; Ethier et al., 1987) a possible autocrine-mediated interaction between these factors may be primarily responsible for the estrogen-induced effects in hormone responsive cells and for the autonomous growth of hormone independent cells. The reduced expression of autocrine growth inhibitory factors, for example TGF-ß, may also contribute to control of cellular proliferation (Knabbe et al., 1987).

CYTOTOXIC DRUG\HORMONAL INTERACTIONS IN BREAST CANCER

Approximately 40% of all tumors expressing both ER

and PGR fail to respond to endocrine manipulation (King et al., 1982; Stewart et al., 1982; Clark & McGuire, 1983) and the majority of those which do respond will eventually become resistant to all endocrine manipulations. Since breast tumors are markedly heterogeneous with respect to the ER content of the cells (Dexter & Calabresi, 1982; van Netten et al., 1985) the inability of endocrine manipulations alone to effectively eradicate hormone responsive breast tumors is not surprising. Endocrine treatments and cytotoxic drugs have different mechanisms of action, therefore, combinations of these modalities may be more effective than either treatment alone. For example, the tendency for ER negative tumor cells to have a more rapid growth rate than cells expressing high levels of ER (Jonat & Maas, 1978; Antoniades & Spector, 1982) may result in cytotoxic drugs killing the ER negative subpopulations while the altered hormonal environment would kill hormone dependent subpopulations.

From the percentage of patients failing to respond to either chemotherapy or endocrine therapy alone, combined chemohormonal treatments would be predicted to reduce the proportion of non-responders from approximately 33% to only 22% (Lippman, 1983). However, these estimations apply to all patients. Greater improvements to combined modalities may be induced in specific subsets of patients. For example, the response rate for ER positive postmenopausal patients could be as high as 84% (Rausch & Kiang, 1988). The estimations of response rates to chemohormonal therapies assume that the combined therapies will produce additive effects and requires that the treatments are non-interactive (Lippman, 1983; Rausch & Kiang, 1988). However, there is considerable experimental evidence to suggest that chemotherapeutic and endocrine manipulative techniques induce interactive processes at the cellular level.

There are a number of possible interactions between hormones and cytotoxic drugs which could influence the overall response to a combined modality. For example, one agent could alter the dynamics of drug absorption, distribution, elimination or metabolism of another and thereby perturb either host toxicity or cytotoxicity. Furthermore, since the majority of antineoplastic agents are cell cycle specific, alterations in the cell cycle distribution of the target tissue could induce significant changes in the cytotoxicity of these drugs.

Additive\synergistic chemohormonal interactions.

Braunschweiger & Schiffer (1980) observed that the most effective sequential combination chemotherapy regimens are those where the drugs are administered coincident with the cell kinetic recovery from single agents. Recovery from one agent is accompanied by an increase in the number of cells in $S-G_2$, thereby producing a cell population more sensitive to the cytotoxic effects of another cell cycle specific agent. The main purpose of many chemohormonal modalities is to utilize endocrine manipulations to produce similarly advantageous perturbations in the cell cycle profile of tumors.

Estrogenic recruitment of hormone responsive breast cancer cells is based on the observations that estrogen stimulates the proliferation of some ER positive breast cancer cells *in vitro* (Lippman et al., 1976; Aitken & Lippman, 1982) and produces an increase in the number of cycling cells as determined by increases in the proportion of cells in S phase (Weichselbaum et al., 1978). When administered prior to antineoplastic drugs, estrogens can enhance the cytotoxicity of a number of chemotherapeutic agents in both *in vitro* and *in vivo* experimental systems. Thus, estradiol potentiates the cytotoxic effects of adriamycin (Hug et al., 1986; Bontenbal et al., 1988), methotrexate (Clarke et al., 1985) cytosine arabinoside (Weichselbaum et al., 1978) and cyclophosphamide (Markaverich et al., 1983; Paridaens et al., 1986).

In order to further increase the proportion of cells entering drug-sensitive phases of the cell cycle, a number of investigators have attempted to use phase specific-inhibition:rescue techniques to synchronize cell populations. Tamoxifen induces a blockade in $G_0\backslash G_1$ (Benz et al., 1983; Taylor et al., 1983) which is reversed by estradiol (Lippman et al., 1976) producing a synchronous cohort of cells which progress into S phase (Aitken & Lippman, 1982). More recently, Yang & Samaan (1987) have reported improved synchronization by treatment with low concentrations of thymidine following blockade and rescue with tamoxifen and estradiol. Furthermore, cell populations synchronized in this manner are at least 50-fold more sensitive to the inhibitory effects of a pharmacologically relevant concentration of 5-fluorouracil.

Alterations in the rate of proliferation may also influence the cellular metabolism of some cytotoxic agents. For example, methotrexate is metabolized to longer polyglutamate derivatives in more rapidly proliferating human breast cancer cells (Kennedy et al., 1985). These longer metabolites are preferentially retained intracellularly and exhibit an equivalent potential for inhibiting dihydrofolate reductase to the parent drug (Kennedy et al., 1983). Thus, they may contribute to the greater cytotoxicity of methotrexate observed in more rapidly proliferating, estrogen-stimulated cells (Clarke et al., 1985).

Perturbations in drug uptake may also influence

cytotoxicity. Bontenbal et al. (1988) reported that estradiol increases both the cellular uptake of adriamycin and its cytotoxicity in MCF-7 cells. The proportion of cells in the S-G$_2$\M phases of the cell cycle are also increased.

Tamoxifen administered alone has been reported to synergistically increase the cytotoxicity of 5-fluorouracil in MCF-7 cells (Benz et al., 1983). The improved cytotoxicity of tamoxifen:5-fluorouracil was attributed to an ability of the triphenylethylene to increase 5-fluorouridine triphosphate incorporation into RNA. The synthesis of dihydrofolate reductase is reduced following treatment with tamoxifen (Levine et al., 1985) and might, therefore, be expected to increase the cytotoxicity of methotrexate. However, tamoxifen alone does not influence the cytotoxicity of methotrexate but increases the potency of a sequential combination of methotrexate and 5-fluorouracil (Benz et al., 1983). The mechanism by which tamoxifen increases sensitivity to some cytotoxic agents remains unclear. Tamoxifen can inhibit calmodulin (Gulino et al., 1986) and trifluoroperazine inhibitors of calmodulin function can increase sensitivity to adriamycin by mechanisms other than alterations in drug accumulation or retention (Ganapathi et al., 1988).

Other hormones in addition to steroids and the triphenylethylene antiestrogens may alter the sensitivity of tumor cells to cytotoxic drugs. For example, insulin increases the cytotoxicity of methotrexate in MCF-7 cells (Alabaster et al., 1981; Clarke et al., 1982). This potentiation may reflect both the ability of insulin to increase the percentage of cells in S phase (Gross et al., 1984) and altered intracellular metabolism of the drug. Insulin induces a 30% increase in the contribution of higher molecular weight polyglutamate metabolites. These methotrexate polyglutamates are retained intracellularly for a longer period of time than the parent drug but are equally cytotoxic (Kennedy et al., 1983).

Endocrine treatment can also improve the therapeutic index of some cytotoxic drugs by altering host toxicity. For example, prednisolone reduces host toxicity and enhances the antitumor effects of nitrogen mustard, melphalan and chlorambucil in rats bearing either the Yoshida sarcoma or the Walker carcinosarcoma (Shepherd & Harrap, 1982). In patients with advanced breast cancer the toxicity of a combination of cyclophosphamide, methotrexate and 5-fluorouracil is reduced by fluoxymesterone (Tormey et al., 1981). In some experimental tumor systems progesterone can reduce the systemic toxicity of chlorambucil (Shepherd & Harrap, 1982).

Effects of cytotoxic drugs on circulating hormone levels.

The most widely reported effects of cytotoxic drugs on the levels of circulating hormones relate to perturbations in ovarian function induced by alkylating agents. When administered as a single agents, both cyclophosphamide (Koyama et al., 1977; Schilsky & Erlichman, 1982) and melphalan (Fisher et al., 1979) induce a potentially reversible failure of ovarian function.

There is no clear evidence that cytotoxic chemotherapy influences either pituitary or adrenal function. Adjuvant chemotherapy does not influence adrenal production of either dehydroepiandrosterone and androstenedione (Rose & Davis, 1977).

The contribution of alterations in the serum concentrations of steroids to the response of breast tumors to chemotherapy is not known. However, response to cytotoxic therapy in premenopausal women does not correlate with symptomatic ovarian dysfunction (Bonnadonna et al., 1977).

Chemohormonal regimens in breast cancer patients.

Initial attempts to improve response to chemotherapy by hormonal manipulation produced encouraging results (Lippman et al., 1984; Allegra, 1983; Paridaens et al., 1985). However, subsequent trials produced less favorable responses (Eisenhauer et al., 1984). Overall, it is not clear that significant improvements have been obtained for overall survival by combined chemohormonal regimens (Davidson & Lippman, 1985; Lippman, 1983; Sertoli et al., 1985; Sheth & Allegra, 1987). Some combinations produce considerable increases in host toxicity (Sheth & Allegra, 1987). However, improvements in endpoints other than overall survival have been observed in specific subsets of patients, for example, adjuvant chemohormonal treatment of postmenopausal patients with ER positive tumors (Sertoli et al., 1985).

The reasons for the poor response rates to combined chemohormonal regimens are not known but may reflect suboptimal scheduling of cytotoxic drugs and hormones, adverse interactions between modalities, or inadequate synchronization of tumor cell populations. For example, not all tumors may respond to attempts to manipulate cell cycle profiles. Brünner et al. (in press) have observed that tamoxifen is capable of blocking MCF-7 cell proliferation in $vitro$ in $G_0\backslash G_1$ of the cell cycle but the tamoxifen-induced regression of MCF-7 tumors in nude mice is the result of increases in cell loss and not alterations in cell cycle distribution. Thus, attempts to confirm adequate synchronization or recruitment by concurrent estimations of either thymidine labelling index, primer dependent alpha-DNA polymerase index or cell cycle distribution, for example on fine needle aspirates, are necessary to identify those tumors most likely to respond to the subsequent chemotherapy. Indeed, when these measurements have been performed, improved response rates have been observed in those tumors which exhibited significant estrogenic recruitment (Conte et al., 1985; Conte et al., 1986).

Adverse chemohormonal interactions.

In addition to inadequate estrogenic recruitment, some endocrine manipulations may reduce the cytotoxicity of a number of drugs by inducing adverse perturbations in either cell cycle kinetics or biochemical and pharmacological parameters. Tamoxifen administered alone has been reported to both increase (Benz et al., 1983) and decrease (Hug et al., 1985) the cytotoxicity of 5-fluorouracil in MCF-7 cells *in vitro*. Tamoxifen reduces the proportion of cells in S phase and increases the proportion in the $G_0 \backslash G_1$ phase of the cell cycle. Therefore, by cell kinetic criteria, tamoxifen would be expected to inhibit the cytotoxicity of 5-fluorouracil. The reason for the disparate observations regarding tamoxifen:5-fluorouracil interactions are unclear but may reflect differences in experimental design. The cytotoxic effects of adriamycin are also reduced by tamoxifen in MCF-7 cells growing *in vitro* (Hug et al., 1985).

Tamoxifen increases melphalan efflux and inhibits influx in MCF-7 cells. Reduced steady-state intracellular concentrations of the drug appear to be responsible for the concurrent reduction in cytotoxicity (Goldenberg & Froese, 1985). Pharmacological concentrations of estrogen reduce the intracellular steady-state levels of methotrexate in both hormone dependent (Clarke et al., 1983) and independent cells (Clarke et al., 1985). However, the cytotoxicity of methotrexate is only reduced in hormone independent cells (Clarke et al., 1985).

The mechanism by which steroids and triphenylethylenes inhibit drug uptake is not known. However, pharmacological concentrations of both estradiol and tamoxifen reduce the membrane fluidity of MCF-7 and MDA-MB-436 human breast cancer cells (Clarke et al., 1987). The high lipophilicity of these agents could result in their predominantly partitioning into the bilipid layer of the cellular membranes. A close association between steroids and plasma membranes exists where the molecules either align themselves with their cyclopentano end inserted into lipid (Wilmer, 1961) or lie along the surface of the phospholipid polar heads (Duval et al., 1983). The effect of such interactions between lipophilic agents and cellular membranes could result in the reduced mobility of membrane-associated transport proteins and a consequent alteration of their functional capabilities (Clarke et al., 1987).

Cytotoxic drug effects on hormone receptor expression.

ER expression is altered following exposure to the cytotoxic agent human interferon-alpha. Concentrations of this agent which do not alter cell proliferation increase ER levels up to 7.2 fold in ZR-75-1 human breast cancer cells (van den Berg et al., 1987). Interferon-alpha also increases detectable ER levels in breast or uterine cell homogenates (Dimitrov et al., 1984). The ability to alter ER expression may be a property of all interferons. Non-inhibitory concentrations of interferon-ß increase ER levels in CG-5 human breast cancer cells (Sica et al., 1987). Increased ER levels were observed in skin metastases of breast cancer patients treated with fibroblast interferon (Pouillart et al., 1982). Furthermore, both interferon-alpha and interferon-gamma increase the sensitivity of MCF-7 cells to the antiproliferative effects of antiestrogens (Kangas et al., 1985; van den Berg et al., 1987).

Agents representative of most of the major classes of antineoplastic drugs inhibit ER expression *in vitro*. For example, adriamycin, methotrexate, 5-fluorouracil, vincristine and melphalan can reduce ER expression in both human breast cancer (Clarke et al., 1986; Muller et al., 1983; Yang & Samaan, 1983) and rat uterine cells (Morris & Stephen, 1983). Whilst ER levels can recover rapidly following only one cycle of cytotoxic treatment *in vitro* (Yang & Samaan, 1983; Clarke et al., 1986) drug resistance induced by prolonged exposure to stepwise increasing concentrations of adriamycin is accompanied by a permanent loss of both ER and PGR expression (Vickers et al., 1988). An important consequence of these alterations in receptor expression is either a markedly reduced response (Clarke et al., 1986) or a complete insensitivity to antiestrogens (Vickers et al., 1988).

The clinical data supporting antineoplastic drug-induced alterations in ER levels in patients is equivocal. Changes from ER positivity to negativity have been reported (Nomura et al., 1985; Toma et al., 1985) but others consider that the majority of tumors may not change their ER status following conventional chemotherapeutic regimens (Allegra et al., 1980; Kiang et al., 1984). These latter observations more closely reflect the ability of cells *in vitro* to recover ER expression rapidly following cytotoxic drug treatment (Clarke et al., 1986). Thus, pretreatment with antineoplastic drugs would be unlikely to alter responses to endocrine manipulations in tumors subsequently regressing from adjuvant or neoadjuvant regimens months or years following cytotoxic chemotherapy. However, cytotoxic drug-induced reductions in ER expression could markedly influence the outcome of chemohormonal treatments where the two modalities are administered either simultaneously or sequentially e.g. in attempts to induce either estrogen recruitment or cell synchronization.

The mechanisms responsible for the drug-induced inhibition of ER expression remain unclear. Some cytotoxic drugs could interact directly with receptor protein e.g melphalan induces cross-linkages in both cytosolic and nuclear proteins (Green et al., 1984). Adriamycin has been reported to competitively inhibit estradiol binding to ER (Di Carlo et al., 1978) but this was not subsequently confirmed (Morris & Stephen, 1983; Muller et al., 1980; Horowitz & McGuire, 1978). Whilst cytotoxic agents inhibit ER expression, the affinity of remaining receptors for ligand is unaltered (Clarke et al., 1986; Morris & Stephen, 1983; Muller et al., 1980; Yang & Samaan, 1983). Thus, significant competitive or allosteric interactions between receptors and antineoplastic drugs seems unlikely. 5-fluorouracil may inhibit ER expression by reducing receptor synthesis but adriamycin, melphalan and vincristine at concentrations which inhibit ER expression do not significantly alter global protein synthesis (Clarke et al., 1986). The ability of adriamycin to intercalate (Waldes & Carter, 1981) and of melphalan to cross link DNA strands (Green et al., 1984) may alter transcription. Thus, some cytotoxic drugs may influence the level of ER expression by perturbing the rate of ER mRNA synthesis.

EGF-R expression may be altered following exposure to cytotoxic drugs. MCF-7 cells selected stepwise with increasing concentrations of adriamycin (MCF-7 Adr^R) lose ER expression but increase the levels of detectable EGF-R. Thus, MCF-7 Adr^R cells express 250,000 EGF-R sites\cell compared with approximately 2,000 sites\cell in the parental cells (Vickers et al., 1988). However, this may more closely reflect the altered ER expression or the expression of the p170 glycoprotein, which confers drug resistance, rather than being due to a direct effect of the cytotoxic drugs. An inverse relationship between low ER and elevated EGF-R levels has been described in primary breast tumors (Sainsbury et al., 1985; Perez et al., 1984; Wrba et al., 1988) and human breast cancer cell lines (Davidson et al., 1987).

FUTURE PROSPECTS

Conventional chemotherapeutic and endocrine

manipulative regimens have failed to produce any significant improvements in the overall survival rate of breast cancer patients. The majority of breast tumors exhibit marked heterogeneity with respect to ER\PGR content. Since chemotherapeutic and hormonal modalities have different modes of action, a combined chemohormonal approach should provide the greatest potential for improvements in response rates. However, combined therapies have been relatively unsuccessful in improving the overall survival rate. This may either reflect an inherent inability of combinations of agents currently available to eradicate the disease or an inappropriate scheduling of these agents.

The rationale for estrogenic recruitment\cell synchronization was developed *in vitro* and utilized for the design of clinical trials without detailed investigation in experimental animal systems. This was largely due to the lack of an appropriate animal model in which to study the responses of human breast tumor tissue. Until recently, the estrogen responsive human breast cancer cell lines available required estrogen for tumor growth in ovariectomized nude mice. Consequently, estrogenic recruitment could only be investigated in tumors which were regressing following removal of estrogens and then re-stimulated. We have described of MCF-7 cells (MIII) which has acquired the ability to form proliferating tumors in ovariectomized athymic nude mice. The cells retain ER expression, grow more quickly in estrogen-supplemented animals and are inhibited by antiestrogens (Clarke et al., in press). Consequently, MIII cells growing in ovariectomized nude mice provide an opportunity to more closely investigate the ability of endocrine manipulations to synchronize tumor cells and to alter their sensitivity to cytotoxic drugs.

Cell cycle recruitment or synchronization with mitogenic growth factors may prove more efficient than utilizing steroids. Since the majority of breast tumors appear to express EGF-R, both ER negative and ER positive subpopulations could be stimulated.

Breast cancer cells secrete a number of mitogenic growth factors that may function as autocrine mediators of cell proliferation. Novel chemotherapeutic agents could be directed against these factors and\or their cellular receptors. The ability of a high proportion of breast tumors cells to express EGF-R may provide a target for EGF-linked cytotoxic agents, antibodies or synthetic peptides which block the biological activity of this receptor. Arteaga et al. (1988) have demonstrated the ability of anti-IGF-I antibodies to induce regression of ER negative MDA-MB-231 tumors growing in nude mice. These antibodies also inhibit hormone dependent MCF-7 cells *in vitro*. Thus, it may be possible to design novel therapies which can effectively inhibit both hormone dependent and independent tumors but without the cell kinetic restrictions and side effects of chemotherapeutic drugs. Indeed, it might be anticipated that only those therapeutic regimens that are directed against the heterogeneous nature of breast tumors will ultimately prove successful in the eradication of the disease.

REFERENCES

Aamdal S, Bormer O, Jorgensen O, Host H, Eilassen G, Kaalhus O, Pihl A (1984). Estrogen receptor and long term prognosis in breast cancer. Cancer 53: 2525-2529.

Aitken SC, Lippman ME (1982). Hormonal regulation of net DNA synthesis in MCF-7 human breast cancer cells in tissue culture. Cancer Res 42: 1727-1735.

Akhtar M, Robinson C, Ali MA, Godwin JT (1983). Secretory carcinoma of the breast. Light and electron microscopic study of three cases with review of the literature. Cancer 51: 2245-2254.

Alabaster O, Vonderhaar BK, Shafie SM (1981). Metabolic modification by insulin enhances methotrexate cytotoxicity in MCF-7 human breast cancer cells. Eur J Cancer Clin Oncol 17: 1223-1228.

Allegra JC, Barlock A, Huff KK, Lippman ME (1980). Changes in multiple or sequential estrogen receptor determinations in breast cancer. Cancer 45: 792-794.

Allegra JC (1983). Methotrexate and 5-fluorouracil following tamoxifen and premarin in advanced breast cancer. Semin Oncol 10: 23-28.

Ankerst J & Jonsson N (1989). Adenovirus type 9-induced tumorigenesis in the rat mammary gland related to sex hormonal state. J Natl Cancer Inst 81: 294-298.

Antoniades K, Spector H (1982). Quantitative estrogen receptor values and growth of carcinoma of the breast before surgical intervention. Cancer 50: 793-796.

Arteaga CL, Tandon AK, Von Hoff DD, Osborne CK (1988). Transforming growth factor beta: potential autocrine growth inhibitor of estrogen receptor negative human breast cancer cells. Cancer Res 48: 3898.

Arteaga CL, Kitten L, Coronado E, Jacobs S, Kull F, Osborne CK (1988). Blockade of the type I somatomedin receptor inhibits growth of estrogen receptor negative human breast cancer cells in athymic mice. Abstract 683 Proceedings of the 70th Annual Meeting of the Endocrine Society pp191.

Balakrishnan A, Yang J, Beattie CW, Gupta TKD, Nandi S (1987). Estrogen receptor in dissociated and cultured human breast fibroadenoma epithelial cells. Cancer Lett 34: 233-242.

Barnes D, Sato G (1978). Growth of human mammary tumor cell line in a serum free medium. Nature 281: 388-389.

Bates SE, Davidson NE, Valverius EM, Dickson RB, Freter CE, Tam JP, Kudlow JE, Lippman ME, Salomon S (1988). Expression of transforming growth factor alpha and its mRNA in human breast cancer: its regulation by estrogen and its possible functional significance. Mol Endocrinol 2: 543-545.

Beatson GT (1896). On the treatment of inoperable cases of carcinoma of the mamma: suggestions from a new method of treatment, with illustrative cases. Lancet ii: 104-107-162-165.

Benz C, Cadman E, Gwin J, Wu T, Amara J, Eisenfeld A, Dannies P (1983). Tamoxifen and 5-fluorouracil in breast cancer: Cytotoxic synergism *in vitro*. Cancer Res 43: 5298-5303.

Blanco G, Alavaikko M, Ojala A, Collan Y, Heikkinen M, Hietanen T, Aine R, Taskinen PJ (1984). Estrogen and progesterone receptors in breast cancer: Relationships to tumor histopathology and survivial of patients. Anticancer Res 4: 383-390.

Bonnadonna G, Rossi A, Valagussa P, Banfi A, Veronesi V (1977). The CMF program for operable breast cancer. Lack of association of disease-free survival with depression of ovarian function. Cancer 44: 847-857.

Bontenbal M, Sonneveld P, Foekens JA, Klijn JGM (1988). Oestradiol enhances doxorubicin uptake and cytotoxicity in human breast cancer cells (MCF-7). Eur J Cancer Clin Oncol 24: 1409-1414.

Bradbeer JW, Kyngdom J (1983). Primary treatment of breast cancer in elderly women with tamoxifen. Clin Oncol 9: 31-34.

Bradley SJ, Garfinkle G, Walker E, Salem R, Chen LB, Steele G (1986). Increased expression of epidermal growth factor receptors in human colon carcinoma cells. Arch Surg 121: 1241.

Braunschweiger PG, Schiffer LM (1980). Cell kinetic-directed sequential chemotherapy with cyclophosphamide and adriamycin in T1699 mammary tumors. Cancer Res 40: 737-743.

Briand P (1983). Hormone-dependent mammary tumors in mice and rats as a model for human breast cancer. Anticancer Res 3: 273-282.

Bronson FH, Dagg CP, Snell GD (1975). " Biology of the laboratory mouse." Reproduction 187-204.

Bronzert DA, Pantazis P, Antoniades HN, Kasid A, Davidson N, Dickson RB, Lippman ME (1987). Synthesis and secretion of platelet-derived growth factor by human breast cancer cell lines. Proc Natl Acad Sci USA 84: 5763-5767.

Brünner N, Bronzert D, Vindelov LL, Rygaard K, Spang-Thomsen M, Lippman ME. Effect of growth and cell cycle kinetics of estradiol and tamoxifen on MCF-7 human breast cancer cells grown *in vitro* and in nude mice. Cancer Res (in press).

Butler WB, Kelsey WH, Goran N (1981). The relationship between estrogen receptors and response rates to cytotoxic chemotherapy in metastatic breast cancer. Cancer Res 41: 82-88.

Campbell FC, Blamey RW, Elston CW, Nicholson RI, Griffiths FC, Haybittle JL (1981). Oestrogen-receptor status and sites of metastasis in breast cancer. Br J Cancer 44: 456-459.

Carter SK (1981). The interpretation of trials: combined hormonal therapy and chemotherapy in disseminated breast cancer. Breast Cancer Res Treat 1: 43-51.

Cattoretti G, Andreola S, Clemente C, D'Amato L, Rilke F (1988). Vimentin and P53 expression on epidermal growth factor receptor-positive oestrogen receptor-negative breast carcinomas. Br J Cancer 57: 353-357.

Clark GM, McGuire WL (1983). Progesterone receptors and human breast cancer. Breast Cancer Res Treat 3: 157-163.

Clark GM, McGuire WL (1988). Steroid receptors and other prognostic factors in primary breast cancer. Semin Oncol 15: 20-25.

Clarke R, Brünner N, Katz D, Glanz P, Dickson RB, Lippman ME, Kern FG (in press). The effects of a constitutive production of TGF-alpha on the growth of MCF-7 human breast cancer cells *in vitro* and *in vivo*. Mol Endocrinol.

Clarke R, Morwood J, van den Berg HW, Nelson J, Murphy RF (1986). Effect of cytotoxic drugs on estrogen receptor expression and response to tamoxifen in MCF-7 cells. Cancer Res 46: 6116-6119.

Clarke R, van den Berg HW, Kennedy DG, Murphy RF (1983). Reduction of the antimetabolic and antiproliferative effects of methotrexate by 17 beta-oestradiol in a human breast carcinoma cell line, MDA-MB-436. Eur J Cancer Clin Oncol 19: 19-24.

Clarke R, van den Berg HW, Kennedy DG, Murphy RF (1985). Oestrogen receptor status and the response of human breast cancer cells to a combination of methotrexate and 17 beta-oestradiol. Br J Cancer 51: 365-369.

Clarke R, van den Berg HW, Murphy RF (1982). Modulation by insulin of the antimetabolic effect of methotrexate on human breast cancer cells. Ir J Med Sci 150: 131.

Clarke R, van den Berg HW, Nelson J, Murphy RF (1987). Pharmacological and suprapharmacological concentrations of both 17 beta oestradiol (E2) and tamoxifen (TAM) reduce the membrane fluidity of MCF-7 and MDA-MB-436 human breast cancer cells. Biochem Soc Trans 15: 243-244.

Clarke R, Brünner N, Katzenellenbogen BS, Thompson EW, Norman MJ, Koppi C, Paik S, Lippman ME, Dickson RB (in press). Progression from hormone dependent to hormone independent growth in MCF-7 human breast cancer cells. Proc Natl Acad Sci USA.

Clarysse A (1985). Hormone-induced tumour flare. Eur J Cancer Clin Oncol 21: 545-547.

Cohen S (1983). The epidermal growth factor. Cancer 51: 1787-1791.

Conte PF, Alama A, Di Marco E, Canavese G, Rosso R, Nicolini A (1986). Cytokinetic parameters of locally advanced human breast cancer treated with diethylstilbestrol and chemotherapy. Bas Appl Histochem 30: 227-231.

Conte PF, Fraschini G, Alama A, Nicolino A, Corsaro E, Canavese G, Rosso R, Drewinko B (1985). Chemotherapy following estrogen-induced expansion of the growth fraction of human breast cancer. Cancer Res 45: 5926-5930.

Cremin M, Clarke R, Nelson J, Murphy RF (1987). The response of human breast cancer cells to glucagon. Biochem Soc Trans 15: 241-242.

Cremin M, Clarke R, Nelson J, Murphy RF (in press). Effect of somatostatin and a synthetic analogue SMS-201-995 (Sandostatin) on human breast cancer cells in culture. Ir J Med Sci.

Cohen ND, Hilf R (1974). Influence of insulin on growth and metabolism of 7,12-dimethylbenz(a)anthracene-induced mammary tumors. Cancer Res 34: 3245-3252.

Cooper AP (1835). "Lectures on the principals and practice of surgery" 8th edition. pp333.

Cullen KJ, Yee D, Paik S, Hampton B, Perdue JF, Lippman ME, Rosen N (1988). Insulin-like growth factor II expression and activity in human breast cancer. Abstract 947 Proceedings of the 29th Annual Meeting of the American Association for Cancer Research pp238.

Dao TL (1972). Ablation therapy for hormone dependent tumors. Annu Rev Med 23: 1-18.

Davidson NE & Lippman ME (1985). Combined therapy in advanced breast cancer. Eur J Cancer Clin Oncol 21: 1123-1126.

Davidson NE, Gelmann EP, Lippman ME, Dickson RB (1987). Epidermal growth factor receptor gene expression in estrogen receptor-positive and negative human breast cancer cell lines. Mol Endocrinol 1: 216-223.

Deeks S, Richards J, Nandi S (1988). Maintenance of normal rat mammary epithelial cells by insulin and insulin-like growth factor 1. Exp Cell Res 174: 448-460.

Defeudis FV, Moreau JP (1986). Studies on somatostatin analogues might lead to new therapies for certain types of cancer. Trends Pharmacol Sci 7: 384-386.

Dehoff MH, Elgin RG, Collier RJ, Clemons DR (1988). Both type I and type II insulin-like growth factor receptor binding increase during lactogenesis in bovine mammary tissue. Endocrinology 122: 2412-2417.

De Leon DD, Bakker B, Wilson DM, Hintz RL, Rosenfeld RG (1988). Demonstration of insulin-like growth factor (IGF-I and -II receptors and binding protein in human breast cancer cell lines. Biochem Biophys Res Comm 152: 398-405.

Dexter DL, Calabresi P (1982). Intraneoplastic diversity. Biochem Biophys Acta 695: 97-112.

Di Carlo F, Reboami C, Centi C, Genazinni E (1978). Changes in the concentration of uterine cytoplasmic oestrogen receptor induced by doxorubicin and methotrexate. J Endocrinol 79: 201-208.

Dickson RB, Huff KK, Spencer EM, Lippman ME (1986). Induction of epidermal growth factor-related polypeptides by 17 beta estradiol in MCF-7 human breast cancer cells. Endocrinology 118: 138-142.

Dickson RB, McManaway ME, Lippman ME (1987). Estrogen-induced factors of breast cancer cells partially replace estrogen to promote tumor growth. Science 232: 1540-1543.

Dilley WG (1971). Relationship of mitosis to alveolar development in the rat mammary gland. J Endocrinol 50: 510-515.

Dimitrov NY, Meyer CJ, Strander H, Einhorn S, Cantell K (1984). Interferon as a modifier of estrogen receptors. Ann Clin Lab Med 14: 32.

Duval D, Durant S, Homo-Delarche F (1983). Non-genomic effects of steroids. Interactions of steroid molecules with membrane structures and functions. Biochem Biophys Acta 737: 409-442.

Eidne KA, Flanagan CA, Millar RP (1985). Gonadotropin-releasing hormone binding sites in human breast carcinoma. Science 229: 989-991.

Eisenhauer EA, Bowman DM, Pritchard KI, Paterson AHG, Ragaz J, Plenderleith I, Geggie PHS, Maxwell I (1984). Tamoxifen and conjugated estrogens (premarin) followed by sequenced methotrexate and 5-FU in refractory advanced breast cancer. Cancer Treat Rep 68: 1421-1422.

Ethier SP, Kudla A, Cundiff KC (1987). Influence of hormone and growth factor interactions on the proliferative potential of normal rat mammary epithelial cells in vitro. J Cell Physiol 132: 161-167.

Fisher B, Sherman B, Rockette H, Redmond C, Margolese R, Fisher ER (1979). 1-Phenyl-alanine mustard (L-PAM) in the management of premenopausal patients with primary breast cancer. Lack of association of disease-free survival with depression of ovarian function. Cancer 44: 847-857.

Fitzpatrick SL, Brightwell J, Wittliff JL, Barrows GH, Schultz GS (1984). Epidermal growth factor binding by breast tumor biopsies and relationship to estrogen receptor and progestin receptor levels. Cancer Res 44: 3448-3453.

Foekens JA, Henkelman MS, Fukkink JF, Blankenstein MA, Klijn JGM (1986). Combined effects of buserelin, estradiol and tamoxifen on the growth of MCF-7 human breast cancer cells in vitro. Biochem Biophys Res Comm 140: 550-556.

Furlanetto RW, DiCarlo JN (1984). Somatomedin C receptors and growth effects in human breast cancer cells maintained in long term culture. Cancer Res 44: 2122-2128.

Furr BJA, Nicholson RI (1982). Use of analogs of LH-RH for treatment of cancer. J Reprod Fertil 64: 529-539.

Ganapathi R, Schmidt H, Grabowski D, Melia M, Ratliff N (1988). Modulation in vitro and in vivo of cytotoxicity but not cellular levels of doxorubicin by the calmodulin inhibitor trifluoperazine is dependent on the level of resistance. Br J Cancer 58: 335-340.

Gespach C, Bawab W, de Cremoux P, Calvo F (1988). Pharmacology,molecular identification and functional characteristics of vasoactive intestinal polypeptide receptors in human breast cancer cells. Cancer Res 48: 5079-5083.

Ghirlanda G, Uccioli L, Perri F, Altomonte L, Bertoli A, Manna R, Frati L, Greco AV (1983). Epidermal growth factor, somatostatin, and psoriasis. Lancet i: 65.

Godolphin W, Elwood JM, Spinelli JJ (1981). Estrogen receptor quantitation and staging as complimentary prognostic indicators in breast cancer. Int J Cancer 28: 677-683.

Goldenberg GJ, Froese EK (1985). Antagonism of the cytocidal activity and uptake of melphalan by tamoxifen in human breast cancer cells in vitro. Biochem Pharmocol 34: 763-770.

Goodman MJ, Bulbrook RD, Moore JW (1988). The distribution of estradiol in the sera of normal Caucasian, Chinese, Filipina, Hawaiian and Japanese women living in Hawaii. Eur J Cancer Clin Oncol 24: 1855-1860.

Green JA, Vistica DT, Young RC, Hamilton TC, Rogan AM, Ozols RF (1984). Potentiation of melphalan cytotoxicity in human ovarian cancer cell lines by glutathione depletion. Cancer Res 44: 5427-5431.

Gross GE, Boldt DH, Osborne CK (1984). Perturbation by insulin of human breast cancer cell cycle kinetics. Cancer Res 44: 3570-3575.

Gulino A, Barrera G, Vacca A, Farina A, Ferretti C, Screpanti I, Dianzani MU, Frati L (1986). Calmodulin antagonism and growth-inhibiting activity of triphenylethylene antiestrogens in MCF-7 human breast cancer cells. Cancer Res 46: 6274-6278.

Harmsen HJ, Porsius AJ (1988). Endocrine therapy of breast cancer. Eur J Cancer Clin Oncol 24: 1099-1116.

Haslem SZ, Levely ML (1985). Estrogen responsiveness of normal mouse mammary cells in primary cell culture: association of mammary fibroblasts with estrogenic regulation of progesterone receptors. Endocrinology 116: 1835-1844.

Herbst AL, Ulfelder H, Poskanzer DC (1971). Adenocarcinoma of the vagina: association of maternal stilbestrol therapy with tumor appearance in young women. N Engl J Med 284: 878-881.

Henderson C, Canellos GP (1980). Cancer of the breast:the past decade. N Engl J Med 302: 1-17.

Henry JA, Nicholson S, Farndon JR, Westley BR, May FEB (1988). Measurement of oestrogen mRNA levels in human breast tumors. Br J Cancer 58: 600-605.

Heuson J-C, Benraad TJ (1983). Biology of breast cancer: receptors; workshop report. Eur J Cancer Clin Oncol 19: 1687-1692.

Heuson J-C, Legros N (1970). Effect of insulin and alloxan

diabetes on growth of rat mammary carcinoma *in vivo*. Eur J Cancer 6: 349-351.

Heuson J-C, Legros N (1972). Influence of insulin deprivation on growth of the 7-12 dimethylbenz(a) anthracene-induced mammary carcinoma in rats subjected to alloxan diabetes and food restriction. Cancer Res 32: 226-232.

Hilf R, Hissin PJ, Shafie SM (1978). Regulatory interrelationships for insulin and estrogen action in mammary tumors. Cancer Res 38: 4076-4085.

Horwitz KB, Costlow ME, McGuire WL (1975). MCF-7: A human breast cancer cell line with estrogen, androgen, progesterone and glucocorticoid receptors. Steroids 26: 785-795.

Horwitz KB, McGuire WL (1978). Actinomycin D prevents nuclear processing of estrogen receptor. J Biol Chem 253: 6319-6322.

Huff KK, Kaufman D, Gabbay KH, Spencer EM, Lippman ME, Dickson RB (1986). Human breast cancer cells secrete an insulin-like growth factor-I-related polypeptide. Cancer Res 46: 4613-4619.

Huff KK, Knabbe C, Lindsey R, Kaufman D, Bronzert D, Lippman ME, Dickson RB (1988). Multihormonal regulation of insulin-like growth factor-I-related protein in MCF-7 human breast cancer cells. Mol Endocrinol 2: 200-208.

Huff KK, McManaway ME, Paik S, Brunner N (1988). *In vivo* effects of Insulin-like growth factor-I (IGF-I) on tumor formation by MCF-7 human breast cancer cells in athymic mice. 70th Annual Meeting Endocrine Society Abstract: 870 238-238.

Hug V, Hortobagyi GN, Drewinko B, Finders M (1985). Tamoxifen-citrate counteracts the antitumor effects of cytotoxic drugs *in vitro*. J Clin Oncol 3: 1672-1677.

Hug V, Johnston D, Finders M, Hortobagyi G (1986). Use of growth-stimulatory hormones to improve the *in vitro* therapeutic index of doxorubicin for human breast tumors. Cancer Res 46: 147-152.

Imai Y, Leung CKH, Freisen HG, Shiu RP (1982). Epidermal growth factor receptors and effect of epidermal growth factor on growth of human breast cancer cells in long term tissue culture. Cancer Res 42: 4394-4398.

Jick H, Walker AM, Rothman KJ (1980). The epidemic of endometrial cancer: a commentary. Am J Pub Health 70: 264-267.

Jonat W, Maat H (1978). Some comments on the necessity of receptor determination in human breast cancer. Cancer Res 38: 4305-4306.

Jones B, Russo J (1987). Influence of steroid hormones on the growth fraction of human breast carcinomas. Am J Clin Pathol 88: 132-138.

Kamby C, Andersen J, Ejlertsen B, Birkler NE, Rytter L, Zedeler K, Thorpe SM, Norgaard T, Rose C (1988). Histological grade and steroid receptor content of primary breast cancer-impact on prognosis and possible modes of action. Br J Cancer 58: 480-486.

Kangas L, Nieminen A-L, Cantell K (1985). Additive and synergistic effects of a novel antiestrogen toremifene (Fc-1157a) and human interferons on estrogen responsive MCF-7 cells *in vitro*. Med Biol 63: 187-190.

Kennedy DG, Clarke R, van den Berg HW, Murphy RF (1983). The kinetics of methotrexate polyglutamate formation and efflux in a human breast cancer cell line (MDA.MB.436): The effect of insulin. Biochem Pharmacol 32: 41-46.

Kennedy DG, van den Berg HW, Clarke R, Murphy RF (1985). The effect of the rate of cell proliferation on the synthesis of methotrexate poly-gamma-glutamates in two human breast cancer cell lines. Biochem Pharmacol 17: 3087-3090.

Kiang DT, Frenning DH, Goldman AJ, Ascensao VF, Kennedy BJ (1978). Estrogen receptor and responses to chemotherapy and hormonal therapy in advanced breast cancer. N Engl J Med 299: 1330-1334.

Kimball ES, Bohn WH, Cockley KD, Warren TC, Sherwin SA (1984). Distinct high-performance liquid chromatography pattern of transforming growth factor activity in the urine of cancer patients as compared with that of normal individuals. Cancer Res 44: 3613-3619.

King RJB, Stewart JF, Millis RR, Rubens RD, Hayward JL (1982). Quantitative comparison of estradiol and progesterone receptor contents of primary and metastatic human breast tumors in relation to response to endocrine treatment. Breast Cancer Res Treat 2: 339-346.

Kinne DW, Ashikari R, Butler A, Menendez-Botet C, Rosen PR, Schwartz M (1981). Estrogen receptor protein in breast cancer as a predictor of recurrence. Cancer 47: 2364-2367.

Knabbe CK, Lippman ME, Wakefield LM, Flanders KC, Kasid A, Derynck R, Dickson RB (1987). Evidence that transforming growth factor-beta is a hormonally regulated negative growth factor in human breast cancer cells. Cell 48: 417-428.

Koga M, Sutherland RL (1987). Epidermal growth factor partially reverses the inhibitory effects of antiestrogens on T47D human breast cancer cell growth. Biochem Biophys Res Comm 146: 738-745.

Koyama H, Wada T, Nishizawa Y, Iwanaga T, Aoki Y, Terasawa T, Kosaki G, Wada A (1977). Cyclophosphamide-induced ovarian failure and its therapeutic significance in patients with breast cancer. Cancer 39: 1403-1409.

L'Hermite M, L'Hermite-Baleriaux M (1988). Prolactin and breast cancer. Eur J Cancer Clin Oncol 24: 955-958.

Lacassagne A (1932). Apparition de cancers de la mamelle chez la souris mâle, soumise à des injections de folliculine. C R Acad Sci (Paris) 195: 639-632.

Lamberts SWJ, Uitterlinden P, Verschoor L, van Dongen KJ, Del Pozo E (1985). Long-term treatment of acromegaly with the somatostatin analogue SMS 201-995. N Engl J Med 313: 1576-1579.

Leake RE, George WD, Godfrey D, Rinaldi F (1988). Regulation of epidermal growth factor receptor synthesis in breast cancer cells. Br J Cancer 58: 521.

Levine RM, Rubalacaba E, Lippman ME, Cowan KH (1985). Effects of estrogen and tamoxifen on the regulation of dihydrofolate reductase gene expression in a human breast cancer cell line. Cancer Res 45: 4681-4690.

Lieberman ME, Maurer RA, Gorski J (1978). Estrogen control of prolactin synthesis *in vitro*. Proc Natl Acad Sci USA 75: 5946-5949.

Lingham RB, Stancel GM, Loose-Mitchel DS (1988). Estrogen regulation of epidermal growth factor messenger ribonucleic acid. Mol Endocrinol 2: 230-235.

Lippman ME (1983). Efforts to combine endocrine therapy and chemotherapy in the management of breast cancer: Do two and two equal three? Breast Cancer Res Treat 3: 117-127.

Lippman ME, Bolan G, Huff K (1976). The effects of estrogens and antiestrogens on hormone responsive human breast cancer cells in long term tissue culture. Cancer Res 36: 4595-4601.

Lippman ME, Cassidy J, Wesley J, Young RC (1984). A randomized attempt to increase the efficacy of cytotoxic chemotherapy in metastatic breast cancer by hormonal synchronization. J Clin Oncol 2: 28-36.

Lyons WR (1958). Hormonal synergism in mammary growth. Proc Royal Soc (Biol) London 149: 303-325.

Markaverich BM, Medina D, Clark JH (1983). Effects of combination estrogen:cyclophosphamide treatment on the growth of the MXT transplantable mammary tumor in the mouse. Cancer Res 43: 3208-3211.

McGrath CM (1983). Augmentation of the response of normal mammary epithelial cells to estradiol by mammary stroma. Cancer Res 43: 1355-1360.

Mehta RG, Banerjee MR (1975). Action of growth promoting hormones on macromolecular biosynthesis during lobuloalveolar development of the entire mammary gland in organ culture. Acta Endocrinol 80: 501-516.

Meyers JS, Rao BR, Stevens SC, White WL (1977). Low incidence of estrogen receptor in breast carcinomas with rapid rates of cellular proliferation. Cancer 40: 2290-2298.

Michnovicz JJ, Hershcopf RJ, Naganuma H, Bradlow HL, Fishman J (1986) Increased 2-hydroxylation of estradiol as a possible mechanism for the anti-estrogenic effect of cigarette smoking. N Engl J Med 315: 1305-1309.

Monaco ME, Lippman ME (1977). Insulin stimulation of fatty acid synthesis in human breast cancer cells in long term tissue culture. Endocrinology 101: 1238-1246.

Moore MR (1980). An insulin effect on cytoplasmic estrogen receptors in the human breast cancer cell line MCF-7. J Biol Chem 256: 3637-3640.

Morris ID, Stephen TM (1983). *In vitro* and *in vivo* interactions of methotrexate and other antimetabolites with the oestrogen high affinity receptors of the rat uterus. Br J Cancer 47: 433-437.

Mukku VR, Stancel GM (1985). Regulation of epidermal growth factor receptor by estrogen. J Biol Chem 260: 9820-9824.

Muller RE, Sheard BE, Traish A, Wotiz HH (1980). Effect of chemotherapeutic agents on the formation of estrogen-receptor complex in human breast tumor cytosol. Cancer Res 40: 2941-2942.

Murphy LC, Murphy LJ, Dubij D, Bell GI, Shiu RPC (1988). Epidermal growth factor gene expression in human breast cancer cells: regulation by progestins. Cancer Res 48: 4555-4560.

Murphy LJ, Sutherland RL, Stead B, Murphy LC, Lazarus L (1986). Progestin regulation of epidermal growth factor receptor in human mammary carcinoma cells. Cancer Res 46: 728-734.

Murr SM, Bradford GE, Geschwind II (1974). Plasma luteinizing hormone, follicle-stimulating hormone and prolactin during pregnancy in the mouse. Endocrinology 94: 112-116.

Myal Y, Shiu R, Bhaumick B, Bala M (1984). Receptor binding and growth-promoting activity of insulin-like growth factors in human breast cancer cells (T47D) in culture. Cancer Res 44: 5486-5490.

Neal DE, Marsh C, Bennet MK, Abel PD, Hall RR, Sainsbury JRC, Harris AL (1985). Epidermal-growth factor receptors in human bladder cancer: comparison of invasive and superficial tumors. Lancet i: 366-368.

Nicholson S, Halcrow P, Sainsbury JRC, Angus B, Chambers P, Farndon JR, Harris AL (1988). Epidermal growth factor receptor (EGFr) status associated with failure of primary endocrine therapy in elderly postmenopausal patients with breast cancer. Br J Cancer 58: 810-814.

Nirmul D, Pegoraro RJ, Naidoo C, Joubert SM (1982). The sex hormone profile of male patients with breast cancer. Br J Cancer 48: 423-427.

Nomura Y, Tashiro H, Shinozuka K (1985). Changes of steroid hormone receptor content by chemotherapy and\or endocrine therapy in advanced breast cancer. Cancer 55: 546-551.

Oka T, Kurachi H, Yoshimura Y, Tsutsumi O, Cossu MF, Tag M (1987). Study of the growth factors for the mammary gland: Epidermal growth factor and mesenchyme-derived growth factor. Nucl Med Biol 14: 353-360.

Osborne CK, Bolan G, Monaco ME, Lippman ME (1976). Hormone responsive human breast cancer in long term tissue culture:effects of insulin. Proc Natl Acad Sci USA 73: 4536-4540.

Osborne CK, Hamilton B, Nover M (1982). Receptor binding and processing of epidermal growth factor by human breast cancer cells. J Clin Endocrinol Metab 55: 86-93.

Osborne CK, Hamilton B, Titus G, Livington RB (1980). Epidermal growth factor stimulation of human breast cancer cells in culture. Cancer Res 40: 3261-2366.

Osborne CK, Monaco ME, Lippman ME, Kahn CR (1978). Correlation among insulin binding, degradation and biological activity in human breast cancer cells in long term tissue culture. Cancer Res 38: 94-102.

Paridaens R, Blonk J, Julien JP, Clarysse A, Ferrazzi E, Rotmensz N, Heuson JC (1985). Aminogluthemide and estrogenic stimulation before chemotherapy for treatment of breast cancer. Preliminary results of a phase II study conducted by the E.O.R.T.C. Breast Cancer Cooperative Group. J Steroid Biochem 23: 1181-1183.

Paridaens R, Klijn JGM, Julien JP, Clarysse A, Rotmensz N, Sylvester R (1986). Chemotherapy with estrogenic recruitment in breast cancer: experimental background and clinical studies conducted by EORTC breast cancer cooperative group. Eur J Cancer Clin Oncol 22: 728.

Patterson J, Furr B, Wakeling A, Battersby L (1982). The biology and physiology of "Nolvadex" (tamoxifen) in the treatment of breast cancer. Breast Cancer Res Treat 2: 363-374.

Pekonen F, Partanen S, Makinen T, Rutanen E-M (1988). Receptors for epidermal growth factor and insulin-like growth factor and their relation to steroid receptors in human breast cancer. Cancer Res 48: 1343-1347.

Peres R, Betsholtz C, Westermark B, Heldin C-H (1987). Frequent expressions of growth factors for mesenchymal cells in human mammary carcinoma cell lines. Cancer Res 47: 3425-3429.

Perez R, Pascual M, Macias A, Lage A (1984). Epidermal growth factor receptors in human breast cancer. Breast Cancer Res Treat 4: 189-193.

Perroteau I, Salomon D, DeBortoli M, Kidwell W, Hazarika P, Pardue R, Dedman J, Tam J (1986). Immunological detection and quantitation of alpha transforming growth factors in human breast carcinoma cells. Breast Cancer Res Treat 7: 201-210.

Petersen OW, Hoyer PE, van Deurs B (1987). Frequency and distribution of estrogen receptor-positive cells in normal, nonlactating human breast tissue. Cancer Res 47: 5748-5751.

Peyrat J-P, Boneterre J, Beuscart R, Djiane J, Demaille A (1988). Insulin-like growth factor I receptors in human breast cancer and their relation to estradiol and progesterone receptors. Cancer Res 48: 6429-6433.

Pilkis SJ, Park CR (1974). Mechanism of action of insulin. Ann Rev Pharmocol 14: 365-388.

Pouillart P, Palangie T, Jouve M, Garcie GE, Fridman WH, Magdalena H, Falcoff E, Billianus A (1982). Administration of fibroblast interferon to patients with advanced breast cancer: possible effects on skin metastasis and on hormone receptors. Eur J Cancer Clin Oncol 18: 929-935.

Rausch D, Kiang DT (1988). Interaction between endocrine and cytotoxic therapy. In Stoll BA (ed): "Endocrine management of cancer. Vol 2" Karger: 102-118.

Raynaud A (1955). Observations sur les modifications provoquees par les hormones oestrogenes, du mode de developpment des mamelons des foetus de souris. C R Acad Sci (Paris) 240: 674-676.

Riss TL, Sirbasku DA (1987). Growth and continuous passage of COMMA-D mouse mammary epithelial cells in hormonally defined serum-free medium. Cancer Res 47: 3776-3782.

Rohlik QT, Adams D, Kull FC, Jacobs S (1987). An antibody to the receptor for insulin-like growth factor I inhibit the growth of MCF-7 cells in tissue culture. Biochem Biophys Res Comm 149: 276-281.

Rose C, Theilade K, Boesen E, Salimtschik M, Dombernowski P, Brünner N, Kjaer M, Mouridsen HT (1982). Treatment of advanced breast cancer with tamoxifen. Breast Cancer Res Treat 2: 395-400.

Rose DP, Davis TE (1980). Effects of adjuvant chemohormonal therapy on the ovarian and adrenal function of breast cancer patients. Cancer Res 40: 4043-4047.

Rose DP, Davis TE (1977). Ovarian function in patients receiving adjuvant chemotherapy for breast cancer. Lancet i: 1174-1176.

Rosenfeld RI, Furlanetto R, Bock D (1983). Relationship of somatomedin-C concentration to pubertal changes. J Pediatrics 103: 723-728.

Sainsbury JRC, Farndon JR, Needham GK, Malcolm AJ, Harris AL (1987). Epidermal growth factor receptor status as predictor of early recurrence and death from breast cancer. Lancet i: 1398-1402.

Sainsbury JRC, Malcolm A, Appleton D, Farndon JR, Harris AL (1985). Presence of epidermal growth factor receptors as an indicator of poor prognosis in patients with breast cancer. J Clin Path 38: 1225-1228.

Sainsbury JRC, Nicholson S, Angus B, Farndon JR, Malcolm AJ, Harris AL (1988). Epidermal growth factor receptor status of histological sub-types of breast cancer. Br J Cancer 58: 458-460.

Sarup JC, Rao KVS, Fox CF (1988). Decreased progesterone binding and attenuated progesterone action in cultured human breast carcinoma cells treated with epidermal growth factor. Cancer Res 48: 5071-5078.

Scambia G, Panici PB, Baiocchi G, Perrone L, Gaggini C, Iacobelli S, Mancuso S (1988). Growth inhibitory effect of LH-RH analogs on human breast cancer cells. Anticancer Res 8: 187-190.

Schlisky RC, Erlichman C (1982). Late complications of chemotherapy: infertility and carcinogenesis. In Chabner B (ed): "Pharmacologic principals of cancer treatment."

Schinzinger A (1889). über carcinoma mammae. Zent Org Gesamte Chir 29: 55-57.

Seibert K, Shafie SM, Triche TJ, Whang-Peng JJ, O'Brien SJ, Toney JH, Huff KK, Lippman ME (1983). Clonal variation of MCF-7 breast cancer cells in vitro and in athymic nude mice. Cancer Res 43: 2223-2239.

Sertoli MR, Scarsi PG, Rosso R (1985). Rationale for combining chemotherapy and hormonal therapy in breast cancer. J Steroid Biochem 23: 1097-1103.

Setyono-Han B, Henkelman MS, Foekens JA, Klijn JGM (1987). Direct inhibitory effects of somatostatin (analogs) on the growth of human breast cancer cells. Cancer Res 47: 1566-1570.

Shafie SM, Brooks SC (1977). Effect of prolactin on growth and estrogen levels of human breast cancer cells (MCF-7). Cancer Res 37: 792-799.

Shafie SM, Grantham FH (1981). Role of hormones in the growth and regression of human breast cancer cells (MCF-7) transplanted into athymic nude mice. J Natl Cancer Inst 67: 51-56.

Shepherd R, Harrap KR (1982). Modulation of the toxicity and antitumor activity of alkylating drugs by steroids. Br J Cancer 45: 413-420.

Sherwin SA, Twardzik DR, Bohn WH, Cockley KD, Todaro GJ (1983). High-molecular-weight transforming growth factor activity in the urine of patients with disseminated cancer. Cancer Res 43: 403-407.

Sheth SP, Allegra JC (1987). What role for concurrent chemohormonal therapy in breast cancer? Oncology 1: 19-24.

Sica G, Natoli V, Stella C, Del Bianco S (1987). Effect of natural beta-interferon on cell proliferation and steroid receptor level in human breast cancer cells. Cancer 60: 2419-2423.

Singh L, Wilson AJ, Baum M, Whimster WF, Birch IH, Jackson IM, Lowrey C, Palmer MK (1988). The relationship between histological grade, oestrogen receptor status, events and survival at 8 years in the NATO (Nolvadex) trial. Br J Cancer 57: 612-614.

Skoog L, Humla S, Axelsson M, Frost M, Norman A, Nordenskjöld B, Waagren A (1987). Estrogen receptor levels and survival of breast cancer patients. Acta Oncol 26: 95-100.

Soule HD, McGrath CM (1980). Estrogen responsive proliferation of clonal human breast carcinoma cells in athymic mice. Cancer Lett 10: 177-189.

Spring-Mills EJ, Stearns SB, Numann PJ, Smith PH (1984). Immunocytochemical localization of insulin- and somatostatin-like material in human breast tumors. Life Sci 35: 185-190.

Stampfer M, Hallowes RC, Hackett AJ (1980). Growth of normal human mammary cells in culture. In vitro 16: 415-425.

Stewart J, King R, Hayward JL, Rubens RD (1982). Estrogen and progesterone receptor: correlation of response rates, site and timing of receptor analysis. Breast Cancer Res Treat 2: 243-250.

Stewart JF, Hayward JL, Rubens RD, King RJB (1982). Estrogen receptor status of advanced breast cancer immediately before chemotherapy does not predict for response. Chemother Pharmacol 9: 124-125.

Stromberg K, Hudgins WR, Dorman LS, Henderson LE, Sowder SC, Sherrell BJ, Mount CD, Orth DN (1987). Human brain tumor-associated urinary high molecular weight transforming growth factor: a high molecular weight form of epidermal growth factor. Cancer Res 47:

Sutton H, Suhrbier K (1967). The estrous cycle and DNA synthesis in the mammary gland. Argonne National Laboratories US Atomic Energy Report pp 157-158.

Swain S, Dickson RB, Lippman ME (1986). Anchorage independent epithelial colony stimulating activity in human breast cancer cell lines. Proceedings 77th Annual Meeting AACR Abstract 844: 213.

Taylor IW, Hodson PJ, Green MD, Sutherland RL (1983). Effects of tamoxifen on cell cycle progression of synchronous MCF-7 human mammary carcinoma cells. Cancer Res 43: 4007-4010.

Terry PM, Ball EM, Ganguly R, Banerjee MR (1975). An indirect radioimmunoassay for mouse casein using 125I-labelled antigen. J Immunol Methods 9: 123-134.

Thomas DB (1984). Do hormones cause breast cancer?. Cancer 53: 595-604.

Toma S, Leonessa F, Paridaens R (1985). The effects of therapy on estrogen receptors in breast cancer. J Steroid Biochem 23: 1105-1109.

Tonelli QJ, Sorof S (1980). Epidermal growth factor requirement for development of cultured mammary gland. Nature 285: 250-252.

Topper YJ, Freeman CS (1980). Multiple hormone interactions in the developmental biology of the mammary gland. Physiol Rev 60: 1049-1106.

Tormey DC, Gelman R, Band PR, Sears M, Bauer M, Arseneau JC, Falkson G (1981). A prospective evaluation of chemohormonal therapy remission maintenance in advanced breast cancer. Breast Cancer Res Treat 1: 111-119.

Tsutsumi O, Tsutsumi A, Oka T (1987). Importance of epidermal growth factor in implantation and growth of mouse mammary tumors in female nude mice. Cancer Res 47: 4651-4653.

Turkington RW (1969). The role of epithelial growth factor in mammary gland development in vitro. Exp Cell Res 57: 79-85.

Turkington RW (1969). Stimulation of mammary cell proliferation by epidermal growth factor in vitro. Cancer Res 29: 1457-1458.

Turkington RW (1971). In Cameron, IL, Padilla, GM, Zimmerman, AM (eds): "Developmental aspects of the cell cycle." Academic Press 315-355.

Twardzik DR, Kimball ES, Sherwin SA, Ranchalis JE, Todaro GJ (1985). Comparison of growth factors functionally related to epidermal growth factor in the urine of normal and tumor-bearing athymic mice. Cancer Res 45: 1934-1939.

Vana J, Murphy GP, Arnoff BL, Baker HW (1979). Survey of primary liver tumors and oral contraceptive use. J Tox

Envir Health 5: 255-273.

van Netten JP, Algard FT, Coy P (1985). Heterogenous estrogen receptor levels detected via multiple microsamples from individual breast cancers. Cancer 56: 2019.

van den Berg HW, Leahey WJ, Lynch M, Clarke R, Nelson J (1987). Recombinant human interferon alpha increases oestrogen receptor expression in human breast cancer cells(ZR-75-1) and sensitizes them to the anti-proliferative effects of tamoxifen. Br J Cancer 55: 255-257.

van der Burg B, Rutterman GR, Blankenstein MA, DeLaat SW, van Zoelen EJJ (1988). Mitogenic stimulation of human breast cancer cells in a growth factor-defined medium: synergistic action of insulin and estrogen. J Cell Physiol 134: 101-108.

Veale D, Marsh C, Ashcroft T, Harris AL (1985). Epidermal growth factor receptors in non-small cell lung cancer. Br J Cancer 52: 441.

Vickers PJ, Dickson RB, Shoemaker R, Cowan KH (1988). A multidrug-resistant MCF-7 human breast cancer cell line which exhibits cross-resistance to antiestrogens and hormone independent tumor growth in vivo. Mol Endocrinol 2: 886-892.

Vignon F, Bouton MM, Rochefort H (1987). Antiestrogens inhibit the mitogenic effect of growth factors on breast cancer cells in the total absence of estrogens. Biochem Biophys Res Comm 146: 1502-1508.

Vonderhaar BK (1988). Regulation of development of the normal mammary gland by hormones and growth factors. In Lippman ME, Dickson RB (eds): "Breast cancer: Cellular and molecular biology." 251-266.

Vuk-Pavlovic S, Bozikof V, Pavelic K (1982). Somatostatin reduced proliferation of murine aplastic carcinoma conditioned to diabetes. Int J Cancer 29: 683-686.

Waldes H, Carter MS (1981). The interaction of adriamycin with nuclear DNA: evidence for a drug induced compaction of isolated chromatin. Biochem Biophys Res Comm 98: 95-101.

Warner MR (1978). Effect of perinatal oestrogen on the pretreatment required for mouse lobular formation in vitro. J Endocrinol 77: 1-10.

Weichselbaum RR, Hellman S, Piro AJ, Nove JJ, Little JB (1978). Proliferation kinetics of a human breast cancer cell line in vitro following treatment with 17 -estradiol and 1-beta-D-arabinofuranosylcytosine. Cancer Res 38: 2339-2342.

Williams G, Howell A, Jones M (1988). The relationship of body weight to response to endocrine therapy, steroid hormone receptors and survival of patients with advanced breast cancer. Br J Cancer 58: 631-634.

Wilmer EN (1961). Steroids and cell surfaces. Biol Rev 36: 368-398.

Wood BG, Washburn LL, Mukherjee AS, Banerjee MR (1975).

Hormonal regulation of lobuloalveolar growth, functional differentiation and regression of whole mouse mammary gland in organ culture. J Endocrinol 65: 1-6.

Wrba F, Reiner A, Ritzinger E, Heinrich H (1988). Expression of epidermal growth factor receptors (EGFR) on breast carcinomas in relation to growth fractions, estrogen receptor status and morphological criteria. Path Res Pract 183: 25-29.

Yang K-P, Samaan NA (1987). Enhancement of hormonal synchronization and 5-fluorouracil cytotoxicity in breast cancer cells by low concentration of thymidine. Anticancer Res 7: 59-64.

Yang KP, Samaan N (1983). Reduction of estrogen receptor concentration in MCF-7 human breast carcinoma cells following exposure to chemotherapeutic drugs. Cancer Res 43: 3534-3538.

Yee D, Cullen KJ, Paik S, Perdue JF, Hampton B, Schwartz A, Lippman ME, Rosen N (1988). Insulin-like growth factor II mRNA expression in human breast cancer. Cancer Res 48: 6691-6696.

Zwiebel JA, Bano M, Nexo E, Salomon DS, Kidwell WR (1986). Partial purification of transforming growth factors from human milk. Cancer Res 46: 933-939.

Molecular Endocrinology and Steroid
Hormone Action, pages 279–293
© 1990 Alan R. Liss, Inc.

LONG-TERM PROGNOSTIC IMPLICATIONS OF SEX-STEROID RECEPTORS IN HUMAN CANCER

K.S. McCarty, Jr., M.D., Ph.D., L.B.Kinsel, M.D,
G.Georgiade, M.D., G.Leight,M.D., and K.S.
McCarty,Sr.,Ph.D.

Department of Pathology (KSM, LBK),Surgery (GG, GL)
and Biochemistry (KSM,Sr), Duke University Medical
Center Durham, North Carolina, 27710

ABSTRACT

Progesterone receptor analysis has been used to
enhance the prognostic usefulness of estrogen receptor
analysis in breast cancer. Immunocytochemical assays for
both steroid receptors have been shown to correlate with
established biochemical techniques but lack long term
clinical follow-up studies to validate their use. One
hundred fifty- two patients were followed for up to 10
years after primary surgical treatment. Steroid receptor
analyses, using both biochemical and immunocytochemical
techniques were performed on their tumor specimens.
Patients with estrogen or progesterone receptor positive
tumors had longer survival than patients with negative
tumors. This difference was most clearly demonstrated
with immunocytochemical analysis of estrogen receptors
($p=0.04$). The two methods for progesterone receptors gave
very similar results. Use of multivariate analysis
revealed that ER by immunocytochemical analysis was the
only significant predictor of prognosis when all four
variables were considered simultaneously ($p=0.04$). This
study suggests that immunocytochemistry gives comparable
results to biochemical analysis for progesterone receptors
but that immunocytochemical analysis of estrogen receptors
was the stronger single prognostic indicator of the four.

INTRODUCTION

Quantitative biochemical estrogen receptor (ER) and

analysis of primary breast cancers has been used as an indicator of survival and disease-free interval (DFI). Progesterone receptor (PgR) analysis began to be used in the late 1970's and early 1980's as an additional prognostic indicator. Recently, immunocytochemical assays for detecting ER and PgR in frozen sections of tissue specimens have become available (1-2). These assays have been shown to correlate with biochemical methods and other prognostic indicators such as histologic and nuclear grade (3-12). ER-ICA has been shown to predict survival and DFI in short and long term clinical follow-up studies (13-15). These new assays offer the advantage of being able to localize steroid receptors within individual tumor components and will hopefully increase the ability to fully characterize steroid receptor expression in heterogeneous tumors.

This study compares estrogen and progesterone receptor expression, measured by both quantitative ligand binding (biochemical) and immunocytochemical techniques with disease-free interval and survival to evaluate the usefulness of immunocytochemical techniques for ER and PgR as prognostic indicators.

PATIENTS AND METHODS:

Patient population: consisted of 152 patients treated at Duke University Medical Center and Cabarrus Memorial Hospital from 1976-1980 for primary breast carcinoma whose tumor specimens were sequentially accessioned for biochemical steroid receptor analysis. Only those patients with sufficient cancerous tissue for immunocytochemical analysis were included in the study. Both pre-and post-menopausal patients were included in the study. Parameters recorded at the time of diagnosis included age, menopausal status, race and stage using the following system:
Stage 1: no positive axillary nodes, no skin or muscle involvement.
2: 1-3 positive nodes, no skin or muscle involvement
3: >=4 positive nodes and/or skin or muscle involvement
4: distant metastases

Parameters recorded during the follow-up period (5-10 years in most cases) included use of adjuvant therapy, date and site of first recurrence, death date or date of last contact without disease.

Ninety-eight of 152 patients received modified radical mastectomies as their primary surgical treatment whereas 30 received radical mastectomies, 8 received simple mastectomies without radiation, 5 received tumor excision without radiation and 11 received tumor excision with radiation. Thirty-nine patients received adjuvant chemotherapy with cytoxan-methotrexate-5FU, alkeran, tamoxifen or other cytotoxic agents (adriamycin or vincristine).

Steroid Receptor Analysis:

Tissue Preparation: Tissues were quick frozen after being washed in 0.005M 4-(2-hydroxy-=ethyl) -1-piperazine ethanesulfonic acid-0.01M Tris-0.0015M EDTA-0.01M thioglycerol-0.02% NaN3 buffer, pH 7.4, and were maintained at -70 degrees Celsius in airtight liquid nitrogen capsules from the time of excision until sectioning for immmunocytochemical analysis.

Quantitative (Biochemical) Methods: Estrogen and progesterone receptor analysis using dextran-coated charcoal and/or sucrose density gradient methods were performed at the time that the tissue was first obtained (16). Results were expressed as femtomoles of estrogen or Organon 2058 binding per mg of cytosol protein. Results greater than or equal to 10 fmol/mg protein were considered positive.
Immunocytochemical Methods: Cryostat sections were fixed in 3.7% formaldehyde-140mM NaCl, buffered with 0.1M phosphate, pH 7.4, at 25 degrees Celsius for 10 minutes, followed by immersion in 100% methanol for 4 minutes and acetone for one minute at -10 degrees Celsius to -25 degrees Celsius. The peroxidase-antiperoxidase method for immunocytochemical localization was performed as previously described (12). Normal goat serum was used as the blocking agent. The primary antibodies H222 (anti-ER) and B39 (anti-PgR) were used at concentrations of 51.0 and 4.13 ug/ml, respectively. The bridging antibody was goat anti-rat immunoglobulin and the PAP complex was of rat origin. Control slides consisted of sections of cancers adjacent to those stained with the primary antibody for which normal rat serum was used in place of the primary antibodies. The slides were treated with poly-L-lysine to improve adhesion of tissue. Localization of

antigen-antibody complexes with peroxidase was developed using diaminobenzidine-H2O2.

Immunocytochemical localization was scored in a semiquantitative fashion incorporating both the intensity and the distribution of specific staining. The evaluations were recorded as percentages of positively stained target cells in each of four intensity categories which were denoted as 0 (no staining), 1+ (weak but detectable above control), 2+ distinct), and 3+ (strong, with minimal light transmission through stained nuclei). For each tissue, a value designated the HSCORE was derived by summing percentages of cells stained at each intensity multiplied by the weighed intensity of staining:

HSCORE= Pi (i+1)

when i=1,2,3 and Pi varies from 1 to 100%. An HSCORE of 75 or greater was considered positive.

Data Analysis: Clinical parameters, biochemical and immunocytochemical assay values were coded separately in a blinded fashion and maintained as independent files until completion of the study. Statistical analysis of biochemical and immunocytochemical values, and clinical data was performed using Pearson regression analysis, the Cox-Mantel test, the Kaplan-Meir Survival estimate and multivariate regression analysis. The sensitivities and specificities were determined using the qualitative biochemical PgR value as the standard. Immunocytochemical assay results were each classified as true positive (TP), true negative (TN), false positive (FP), or false negative (FN) in relation to the biochemical value. The formula used for sensitivity, specificity and accuracy are as follows (17):

$$Sensitivity=TP/(TP+FN)$$
$$Specificity=TN/(TN+FP)$$
$$Accuracy=(TP+TN)/(TP+TN+FP+FN)$$

The threshold HSCORE for the ER immunocytochemical assay (>75 being categorized as positive) was determined by optimizing sensitivity, specificity and accuracy when compared to the biochemical value in a learning population. This was then verified using a test population (4). The threshold for the PgR-ICA method (>75 being categorized as positive) was originally assigned

using the value determined for the ER assay since they use
the same reagents except for the primary antibody. This
value is examined in this study. The threshold used for
both biochemical assays is the commonly used value of 10
fmol/mg protein.

Results

  Immunocytochemical staining of ER and PgR was
localized in the nucleus of cancer cells. Negative
control slides had minimal background staining.
Intraobserver reliability was r=0.91 for ER and r=0.90 for
PgR whereas interobserver reliability was r=0.91 for ER
and r=0.84 for PgR. Correlations were observed between
the log of the biochemical ER value and ER-ICA HSCORE
(r=0.69, p>0.0001) and between the log of the biochemical
PgR value and PgR=ICA HSCORE (r=0.57, p<0.001). When
sensitivity, specificity and accuracy were determined by
comparing the ICA values with the biochemical values
(using the thresholds of HSCORE>=75) being positive and
biochemical >=10 fmol/mg protein being positive), there
was an overall agreement of 72% (accuracy) for ER and 80%
for PgR (Table 1). There was no relationship between ER
or PgR, determined by either method and tumor cellularity
(ER-ICA r=0.10, log biochemical ER r=0.001, PgR-ICA
r=0.02, log biochemical PgR r=0.07).
  Figures 1A-D show the relationship between steroid
receptor expression and survival using threshold values of
75 HSCORE for the immunocytochemical assays and 10 fmol/mg
protein for the biochemical assays. Patients with
positive PgR values by either ICA or biochemical methods,
have some survival advantage through the entire 10 year
follow-up period as compared to patients with negative PgR
values. The difference in survival decreased after 7
years using either method. The separation follows a
similar pattern between the two methods. The separation
between groups is less distinct using biochemical methods
for ER when compared to PgR or to ER measured by ICA.
ER-ICA gives the best separation between groups of the
four parameters (p=0.04). These relationships were
confirmed with multivariate analysis, examining the
relative contribution of each parameter for predicting
survival when all four were considered simultaneously.
The only parameter to show a significant relationship with
survival was ER by immunocytochemical analysis (Table 2).
  Prediction of DFI using the above thresholds was

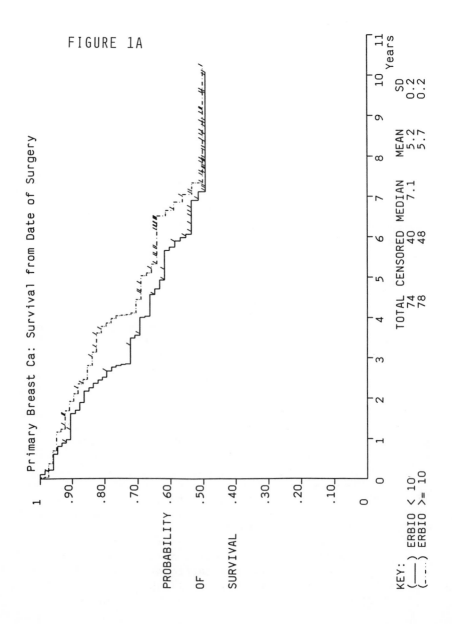

FIGURE 1A

Primary Breast Ca: Survival from Date of Surgery

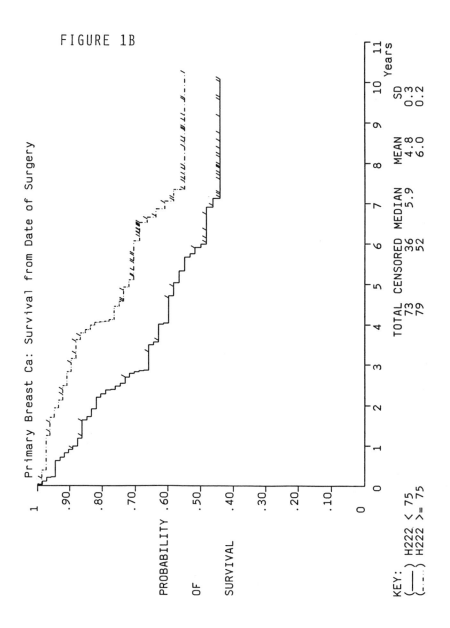

FIGURE 1B

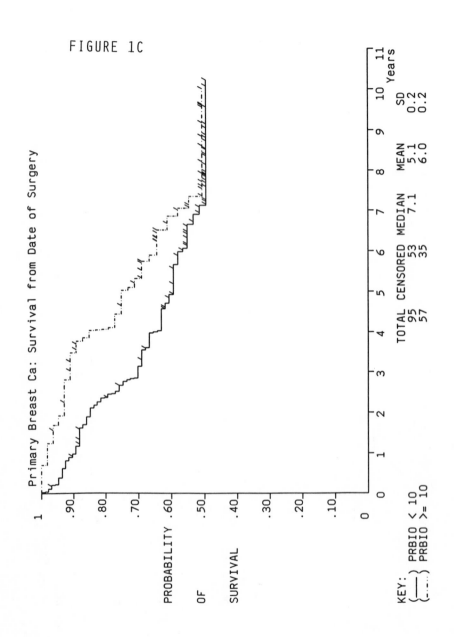

FIGURE 1C

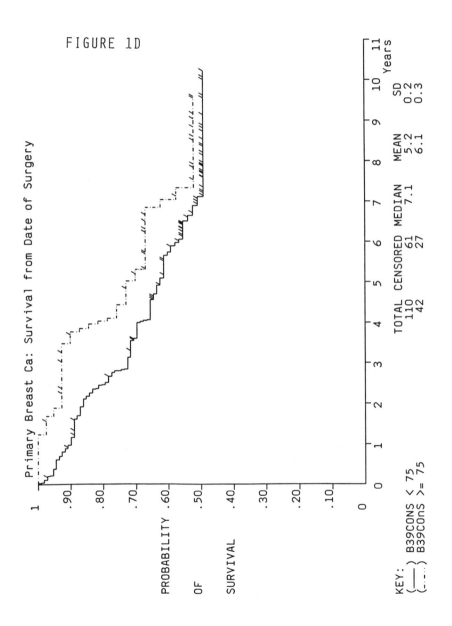

FIGURE 1D

Primary Breast Ca: Survival from Date of Surgery

similar for PgR by biochemical and ICA methods.
Initially, there are longer DFI's in patients with PgR
positive tumors than patients with PgR negative tumors.
After 3 years, however, there was no significant
separation between PgR groups. The curves for ER by either
method shown similar results with the loss of separation
between groups for ICA occuring after 5 years while ER
measured by biochemical methods occurred after 3 years
using these thresholds.

When ER and PgR were used together, results again
vary depending on the techniques used. Using the ligand
techniques, those groups with positive PgR values had
improved survival in the first six years over groups with
negative PgR values, regardless of ER status. Groups
could not be distinguished with either parameter after six
years. The addition of ER to the PgR value did not
identify additional prognostic groups. Using
immunocytochemical techniques, ER detection using antibody
H222 identified a survival advantage in ER positive
groups. PgR detection using B39 further distinguished
subgroups of ER positive tumors, with ER+PgR+ tumors
having a slightly better prognosis than ER+PgR-tumors. A
small groups of ER-PgR+ tumors were observed with an
overall prognosis similar to the ER-PgR-group. These
groups could be distinguished from one another out to 10
years of follow-up.

Discussion

Progesterone receptors, measured with ligand binding
techniques have been shown to correlate with prognosis
(follow-up of 3 years or less) (18-23). The present study
shows survival outcome up to 10 years. Progesterone
receptor was found in previous studies to distinguish
between two prognostic groups using DFI (18-22) and using
survival (20). Sutton, et al., however, found that PgR
did not distinguish between prognostic groups but that ER
did, using both DFI and survival in a three year follow-up
study of Stage I and II patients (23). Progesterone
receptor was found in our laboratory to predict prognosis
out to 7 years using survival and 3 years using DFI but
that after these time periods the differences became
insignificant.

Immunocytochemical analysis (ICA) has been
introduced as an additional method of detecting PgR. It
correlates with biochemical analysis but this relationship

is far from perfect (2). ICA provides a different
perspective to PgR analysis in that it is able to
determine the distribution of receptors in heterogeneous
tissues and assure that the receptors being evaluated
reside in the malignant tissue and not benign or normal
structures. In this study, it was found to have similar
prognostic usefulness to biochemical analysis of PgR.
Positive PgR correlated with a survival advantage out to 7
years and with DFI advantage out to 4 years. When used
independently as a semiquantitative score of the tissue as
a whole, it does not appear to give additional information
over biochemical analysis. However, when used in
combination with analyses of ER using antibody H222, it
was able to define additional prognostic subgroups,
similar to what was observed in studies using biochemical
analysis (18-22).

One reason for the lesser significance in
performance of the comparative assays, may relate to the
threshold value used to assign tumors to ER positive or
negative categories. In studies using biochemical
analysis of ER and PgR, various thresholds have been used
to define receptor positive vs. negative tumors. These
thresholds have been determined using the often subjective
measure of response to hormone therapy (24-25).
Thresholds used for prognostic studies for ER have varied
from 3 to 10 fmol/mg protein and PgR has varied from 1 to
10 fmol/mg protein. The thresholds for ICA analyses of ER
and PgR have been determined by maximizing by agreement
with biochemical analysis (2,4). However, neither method
is without problems and neither should be considered an
optimal standard (2). Survival, in contrast, when defined
specifically as those patients dying of breast cancer (all
other deaths being censored) is an objective measure of
outcome and biologic behavior of the tumor. It should be
used to verify thresholds for ER and PgR analysis using
either biochemical and ICA analysis. Such further testing
of thresholds "steroid receptor" will more effectively
evaluate for the ability of ICA to complement biochemical
analysis of steroid receptors in predicting the biologic
behavior of human tumors.

Table 1. Sensitivity, Specificity and Accuracy of Immunocytochemical Analysis of Estrogen and Progesterone Receptors When Compared to Biochemical Analysis**

	Sen.	Spec.	Accuracy
ER*	59%	86%	72%
PgR*	81%	79%	80%

*ER=Estrogen receptors; PgR= Progesterone receptors
**Threshold positive ligand-binding assay > 10 fmol/mg protein;
Threshold positive immunocytochemical assay > 75

Table 2. Multivariate Regression Analysis of Steroid Receptor Expression and Survival

	Chi Square	p Value
Biochemical ER*	0.374	0.548
Biochemical PgR**	0.992	0.320
ER-ICA***	4.036	0.042
PgR-ICA****	0.365	0.553

*ER=Estrogen Receptors
**PgR=Progesterone Receptors
***ER-ICA=Estrogen Receptors by Immunocytochemical Analysis
****PgR-ICA=Progesterone Receptors by Immunocytochemical Analysis

References

1.   King, W.J., and Greene, G.L.: Monoclonal Antibodies Localize Estrogen Receptor in the Nuclei of Target Cells, Nature 1984; 307: 745-749.

2.   Kinsel, L.B., Flowers, J.L., Cox, E.B., Leight, G.S., Greene, G.L., Konrath, J. and McCarty, K.S., Jr.: Use of a Monoclonal Anti-Progesterone Receptor Antibody to Complement Immunohistochemical Evaluation of ER in Breast Tumors. Lab. Invest. 56: 38, 1987.

3.   McCarty, K.S., Jr., Miller, L.S., Cox, E.B., Konrath, J., and McCarty, K.S., Sr.: Estrogen Receptor Analysis. Correlation of Biochemcial and Immunohistochemical Methods Using Monoclonal Antireceptor Antibodies, Arch. Path. Lab. Med. 1985; 109: 716-721.

4.   McCarty, K.S., Jr., Szabo, E., Flowers, J.L., et al.: Use of a Monoclonal Anti-Estrogen Receptor Antibody in the Immunohistochemical Evaluation of Human Tumors, Cancer Research (Suppl.) 1986; 46: 4244-4248s.

5.   King, W.J., DeSombre, E.R., Jensen, E.G., and Green, G.L., Comparison of Immunocytochemical and Steroid-binding Assays for Estrogen Receptor in Human Breast Tumors, Cancer Research 1985; 45:293-300.

6.   Hawkins, R.A., Sangster, K., and Krajewski, A.: Histochemical Detection of Oestrogen Receptors in Breast Carcinoma: A successful Technique, Brs. J. Cancer 1986; 53: 407-410.

7.   Charpin, C., Martin, P.M., Lissitzky, J. C., et al: Estrogen Receptor Immunocytochemical Assay (ERICA) and Laminin Detection in 130 Breast Carcinomas and Computerized (Samba 200) Multiparametric Quantitative Analysis on Tissue Sections, Bull. Cancer (Paris) 1986; 73: 651-664.

8.   Heubner, A., Beck, T., Greill, H.J., and Pollow, K.: Comparison of Immunocytochemical Estrogen Receptor Enzyme Immunoassay, and Radioligand-labeled Estrogen Receptor Assay in Human Breast Cancer and Uterine Tissue, Cancer Research (Suppl., 1986; 46: 4291s-4295s.

9.   DiFronzo, G., Clemente, C., Capelletti, Y., et al.: Relationship between ER-ICA and Conventional Steroid Receptor Assays in Human Breast Cancer, Breast Cancer Res. Treatment 1986; 8: 35-43.

10.   Brehler, R., Bergholz, M., Rauschecker, H., Blossey, H.C., and Schauer, A.: Immunohistochemische Untersuchungen zur Bestimmung des Hormonrezeptorstatus von

Mammakarzinomen, Klin. Wochenschrift 1986; 64: 370-374.

11. Shintaku, I.P., and Said, J.W.: Detection of Estrogen Receptors with Monoclonal Antibodies in Routinely Processed Formalin-Fixed Paraffin Sections of Breast Carcinoma. Use of DNase Pretreatment to Enhance Sensitivity of the Reaction, Amer. J. Clin. Path. 1987; 87: 161-167.

12. Kinsel, L.B., Flowers, J.L., Szabo, E., et al.: Comparison of Immunocytochemical Analysis of Estrogen and Progesterone Receptors with Histologic Type, Histologic Grade, Nuclear Grade and Quantitative Steroid Receptor Analysis, Amer. J. Path.

13. Thorpe, S.M., Rose, C., Rasmussen, B.B., et al.: Steroid Hormone Receptors as Prognostic Indicators in Primary Breast Cancer, Breast Cancer Res. and Treat. (Suppl) 1986; 7: 91-98.

14. DeSombre, E.R., Thorpe, S.M., Rose, C., et al.: Prognostic Usefulness of Estrogen Receptor Immunocytochemical Assays for Human Breast Cancer, Cancer Research (Suppl) 1986; 46: 4256s-4264s.

15. Kinsel, L.B., Flowers, J.L., Szabo, E., Greene, G.L., Konrath, J., and McCarty, K.S., Jr.: Immunocytochemical Analysis of Estrogen Receptors as a Predictor of Prognosis in Breast Cancer Patients. Comparison with Quantitative Biochemical Methods, Cancer Research.

16. McCarty, K.S., Jr., Barton, T.K., Fetter, B.F., Creasman, W.T., and McCarty, K.S., Sr.: Correlation of Estrogen and Progesterone Receptors with Histologic Differentiation in Endometrial Adenocarcinoma. Am. J. Pathol. 1979, 96: 171-183.

17. Galen, R.S., and Gambino, S.R.: Beyond Normality: The Predictive Value and Efficiency of Medical Diagnosis, New York, N. Y., John Wiley and Sons, Inc. 1975.

18. Pichon, M.F., Palud, C., Brunet, M.,a and Milgrom, E.R.: Relationship of Presence of Progesterone Receptors to Prognosis in Early Breast Cancer, 1980; 40: 3357-3360.

19. Clark, G.M., McGuire, W.L., Hubay, C.A., Pearson, O.H., and Marshall, J.S.: Progesterone Receptors as a Prognostic Factor in Stage II Breast cancer, NEJM 1983: 309: 1343-1347.

20. Mason, B.H., Holdaway, I.M., Mullins, P.R., Yee, L.H., and Kay, R.G.: Progesterone and Estrogen Receptors as Prognostic Variables in Breast Cancer, Cancer Research 1983; 43: 2985-2990.

21. DiFronzo, G., Cappelletti, Y., Coradini, D., Ronchi, E., and Scavone, G.: Prognostic Significance of

Progesterone Receptors Alone or in Association with Estrogen Receptors in Human Breast Cancer, Tumor 1984; 70: 159–164.

22. Alanko, A., Heinonen, E., Scheinin, T., Tolppanen, E.M., and Yihko, R.: Significance of Estrogen and Progesterone Receptors, Disease-Free Interval, and Site of First Metastasis on Survival of Breast Cancer Patients, Cancer 1985; 56: 1696–1700.

23. Sutton, R., Campbell, M., Cooke, T., Nicholson, R., Griffiths, K., and Taylor, I.: Predictive Power of Progesterone Receptor Status in Early Breast Carcinoma. Brst. J. Surg. 1987; 74: 223–226.

24. Walt, A.J., Singhakowinta, A., Brooks, S.C., and Cortez, A.: The Surgical Implications of Estrophile Protein Estimations in Carcinoma of the Breast, Surgery 1976; 80: 506–512.

25. McCarty, K.S., Jr., Cox., C., Silva, J.S., Woodard, B.H., Mossler, J.A., Haagensen, D.E., Barton, T.K., McCarty, K.S., Sr., and Wells, S.A.: Comparison of Sex Steroid Receptor Analyses and Carcinoembryonic Antigen with Response to Hormone Therapy. Cancer 1980; 46: 2846–2850.

Molecular Endocrinology and Steroid
Hormone Action, pages 295–309
© 1990 Alan R. Liss, Inc.

RECEPTOR-DIRECTED RADIOTHERAPY: A NEW APPROACH TO THERAPY OF STEROID RECEPTOR POSITIVE CANCERS

Eugene R. DeSombre, Alun Hughes, S. John Gatley, Jeffrey L. Schwartz and Paul V. Harper.

Ben May Institute (ERD, AH) and Departments of Radiology (SJG, PVH) and Radiation Oncology (JLS), University of Chicago, Chicago, Illinois 60637

INTRODUCTION

While endocrine therapies have been used for treatment of breast cancer for nearly a century (Beatson 1896), and the increased specificity resulting from the use of the estrogen receptor, ER, (McGuire et al., 1975; DeSombre et al., 1979) and the progestin receptor, PR, (Horwitz et al., 1975) status of the cancer has allowed more effective application of such treatments (DeSombre et al., 1987), not all ER+, Table 1, or even ER+/PR+ cancers respond to endocrine therapy. Certain ER+ cancers, like ovarian carcino-

TABLE 1. Estrogen Receptor Positive Cancers

Cancer Site	% ER +	Overall Remission Rate to Endocrine Therapy
Breast	55–80%	25–35%
Endometrium	40–80%	20–30%
Ovary	20–50%	<5%

ma, show little response to endocrine therapies (DeSombre et al., 1987). Even breast and endometrial cancers, which have higher response rates to such therapies, often recur despite initial response. Part of the reason for this may relate to the fact that current forms of endocrine therapies are cytostatic rather than cytotoxic. Although in some cases initial remission of the cancer can lead to apparent cures, a large portion of patients with advanced disease die of their cancer. Studies on endocrine regulation show that removal of the sources of hormones, or antagonism of their effects, leads to dramatic reduction in the size of normal hormone-dependent tissues. Such tissues remain in an atrophic state until increased levels of hormones restimulate growth. Such is likely be the case for cancers as well. While immune processes can help reduce or eliminate a small aggregate of cancer cells remaining after regression of a tumor, there is considerable attraction to the use of a cytotoxic endocrine therapy to attempt to kill the cancer cells. One such approach is to use the steroid receptor to direct therapy to steroid receptor positive (SR+) cancer cells.

Earlier attempts at estrogen receptor directed therapy were by attaching reactive or toxic groups to steroids in the hope that the steroid would direct the toxin or reactive function to SR+ cancer cells. However, the addition of a fairly large, reactive chemical group usually substantially reduced the affinity of the resulting steroid conjugate for the receptor so that their clinical usefulness has been limited (Everson et al., 1973). We have proposed to use this approach by attaching an Auger electron emitting halogen to an estrogen (Seevers et al., 1986). There is good evidence that both bromine (Katzenellenbogen et al., 1981, 1982; McElvany et al., 1982) and iodine (McElvany et al., 1982; Hochberg and Rosner, 1982; Hanson et al., 1982; Jagoda et al.,1984) can be attached to a number of estrogens in a manner which maintains high affinity for the estrogen receptor. Although most of the current interest in such compounds relates to the use of γ-emitting, ^{123}I-labeled estrogens for the imaging and diagnosis of breast cancer, the use of Auger electron emitting halogens linked to estrogens can provide a much needed new treatment approach for ER+ cancers.

Unlike ß decay which is the emission of a single nuclear electron per disintegration, the Auger electron decay process results in the emission of multiple valence electrons, Fig 1. The process is initiated by the capture of a valence electron by the nucleus or internal conversion (Fig 1.), both processes effecting an electron vacancy in an inner valence shell. This vacancy has a cascade effect. The initial vacancy is filled by an electron from a higher energy level, resulting in a high energy state. This leads to the emission of either a photon or another valence electron of a discrete energy, which for the Auger electron corresponds to the energy difference between the initial

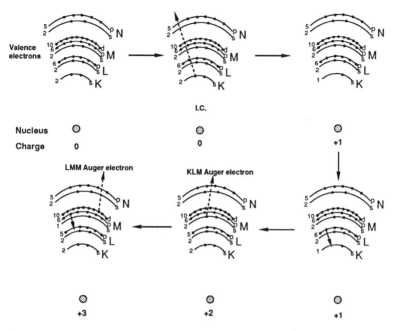

Figure 1. Representation of an Auger decay scheme of bromine-80m. The decay is initiated by internal conversion, represented as the emission of a K-shell electron (middle top), resulting in a +1 charge on the atom (top right). The subsequent filling of the K-shell vacancy (bottom right) results in the emission of a KLM Auger electron (ie. M electron emitted following the L to K shell transition), giving an atom with a +2 charge (middle bottom). The cascade continues, as represented by an M to L shell transition and emission of an LMM Auger electron (bottom left).

and final states of the electron which changed levels minus the binding energy of the emitted electron. The process of filling the initial vacancy therefore not only results in the emission of an electron but the creation of a second valence shell vacancy as well. As this process continues and moves to the outer shells a greater number of electrons are emitted, in effect a shower of electrons, the number and energy distribution of which is characteristic for the atom involved. For bromine-80m, which we will be discussing subsequently and which is illustrated in Fig 1, the Auger process results in the emission of an average of 8 electrons in the condensed state (Powell et al., 1987). What is biologically important is that the energy of these emitted electrons is very low, typically under 10 keV, so that the effective range is in the nanometer range. This

range is ideal for limiting the effectiveness of the subsequent irradiation to a small volume such as the nucleus of
target cells. Kassis, Adelstein, Bloomer and coworkers
(1987) as well as many others have provided a consistent
body of evidence showing the highly effective radiotoxicity
of Auger electron emissions from iodine-125 when this nuclide is actually incorporated into DNA. They have estimated that the shower of the 14 low energy Auger electrons
from iodine-125 over a sphere of 1-5 nm radius is equivalent to a dose of 1.3×10^9 rads/decay.

Theoretically, this situation would seem ideal for treating cancer cells where there is a mechanism to direct the
Auger electron-emitting nuclide to the nucleus, such as
would be expected for a steroid receptor-directed ligand.
The theoretical feasibility of this approach was in fact
established by Bronzert, Hochberg and Lippman (1982) using
$[^{125}I]$ 16α-iodoestradiol (^{125}I-E2) with MCF-7, ER+ breast
cancer cells in culture. They showed that cells treated
with ^{125}I-E2, and frozen down to accumulate sufficient decays, showed a time-dependent reduction in plating efficiency when compared with similar cells treated with ^{125}I-
E2 along with saturating amounts of unlabeled E2 to prevent
the association of the ^{125}I-E2 with ER. Bloomer et al.
(1980) presented compatible results with ^{125}I-tamoxifen,
although they did not use competition with unlabeled estrogen to show unequivocally that the cytoxicity was mediated
by ER.

As important as these reports are with regard to esta-
blishing the theoretical possiblity of using Auger electron
emitting, estrogen receptor-directed ligands for therapy,
it is evident that iodine-125 is not a practical nuclide to
use for such an approach. Obviously one can not "freeze
down" a patient to allow sufficient time for the I-125 to
decay as one can do for cultured cells so the 60 day half
life of this nuclide for practical reasons precludes its
clinical use in this way. However, Br-80m, which decays by
emission of Auger electrons (Fig 1) has a 4.4 hr half life,
which is comparable to the biological half life of the estrogen receptor in vivo (Nardulli and Katzenellenbogen
1986). To establish the applicability of bromine-80m for
this approach it is necessary to actually demonstrate the
radiotoxicity of this nuclide, determine its mean lethal
dose in the nucleus to discover whether the number of atoms
of the nuclide needed for killing is compatible with the

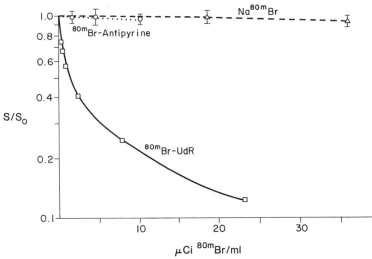

Figure 2. Cytoxicity of bromine-80m in MCF-7 cells. MCF-7 cells plated in 100 mm dishes 24 hr earlier were exposed to [80m]Br-UdR (solid line), [80m]Br-bromoantipyrine (dotted line) or [80m]Br as sodium bromide (dashed line) in MEM, 5% fetal calf serum, pen/strep, and 2 µg/ml insulin, at the concentrations of bromine-80m indicated, for 16 hr. The media were then replaced with the control medium lacking the nuclide and the cells grown for 14 days, rinsed, fixed and stained with crystal violet for colony assays. Colonies of more than 50 cells were counted and the results presented as the ratio of the mean number of colonies in experimental to control conditions. From DeSombre et al., 1988c.

number of ER found in ER+ cancer cells, and show that bromine-80m labeled estrogens are specifically taken up by ER+ tissues in sufficient numbers in vivo to effect such radiotoxicity.

RADIOTOXICITY OF BROMINE-80m

To establish the radiotoxic potential of the nuclide bromine-80m we compared the killing of MCF-7, human breast cancer cells by several chemical forms of the nuclide (Fig 2). When MCF-7 cells were exposed for 16 hr to the nuclide in the form of bromine-80m bromodeoxyuridine, [80m]Br-UdR, which is incorporated into DNA, there was an effective killing as seen by colony assays (Fig 2). Incubation of MCF-7 cells for similar times and with similar concentrations of bromine-80m as either sodium bromide, which does not enter living cells, or bromoantipyrine, which distributes equally in cell water, showed little cytotoxicity (Fig 2). While the initial slope of the survival curve was seen to be linear, characteristic of a high linear energy (LET) effect expected for Auger electrons, the short half life of the bromine-80m, combined with the long doubling time (36 hr) of the MCF-7 cells made more definite estimates of mean lethal dose complicated in this cell system. Therefore, we carried out similar studies with the AA8 subline of Chinese Hamster Ovary (CHO) cells, which have a doubling time of about 16 hr. Initial studies (DeSombre et al., 1988c) showed that an 18 hr incubation of CHO cells with 3.5 μCi/ml ^{80m}Br-UdR reduced the survival to less than 10^{-3}. When incubated for a 2 hour period with a range of doses of ^{80m}Br-UdR a complex survival curve was seen, showing an apparent inital plateau at about 70% survival, in agreement with the 29% labeling index found by autoradiographic analysis of cells fixed after the 2 hr incubation with ^{80m}Br-UdR. The continued decrease in survival seen with increasing doses of ^{80m}Br-UdR, which was inhibitable when unlabeled thymidine was added to the incubations, is believed to result from the incorporation of bromine-80m into the nucleotide pool during the first 2 hr, and continued incorporation into DNA at decreasing efficiency with time after removal of the ^{80m}Br-UdR from the medium.

To circumvent the problem of the low labeling index, we attempted to synchronize CHO cells so that a more substantial proportion of the cells would be in S phase during a 2 hr exposure to ^{80m}Br-UdR. As seen in Figure 3. the survival curve for the synchronized cells showed a linear initial slope, and a plateau at 20-25% survival, consistent with the approximate 75% labeling index found for these cells. When the mean lethal dose (37% survival dose) was estimated, it was found to be about 40 femtocuries per cell, which corresponds to about 45 atoms of bromine per cell (DeSombre et al., 1988c). Considering that the ER content of ER+ cancers ranges from several hundred to about 20,000 sites per cell, the mean lethal dose for bromine-80m shown above is compatible with its proposed function as the nuclide to be attached to the estrogen receptor-directed ligand. Even if there were a small reduction in efficacy of the Auger electrons due to the proximity to, rather than incorporation into, DNA expected for the steroid receptor bound nuclide, one might expect uptake of enough bromoestrogens to cause cytotoxicity in ER+ cancers.

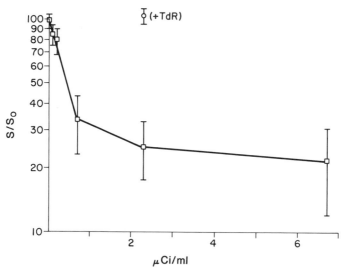

Figure 3. Survival of synchronized CHO cells treated with
80mBr-UdR. CHO cells were synchronized by mitotic shake
off, held at the G1/S border by incubating with 2 μg/ml
aphidicolin, and released from the cell cycle block by re-
plating in fresh medium about 4 hr prior to incubation with
80mBr-UdR, specific activity 1180 Ci/mmol for 2 hr.
100,000 cells, similarly exposed, were removed directly after the 2 hr incubation, washed with PBS, sonicated, pre-
cipitated in cold 10% TCA, washed and counted to determine
the incorporation per cell. For colony assay, the media
were changed in the dishes, the colonies allowed to grow
for 8 days and counted. The labeling index was determined
on cells plated on slide cultures and similarly exposed to
80mBr-UdR. After the 2 hr incorporation period, the slides
were washed, fixed, dipped in Kodak NTB 3 emulsion at 38°
in the dark, exposed overnight and developed. The labeling
index was found to be about 75%.

   It was also of interest to discover the nature of the
damage effected by the incorporation of 80mBr-UdR into the
cells. To that end chromosomal aberrations were studied at
various times after exposure of the CHO cells to 80mBr-UdR.
   The timing of the peak of chromosomal aberrations was
identical to the peak of S phase (as determined by labeling
of cells with 3H thymidine) and most of the aberrations
were of the chromatid type, suggesting that the chromosome
damage was induced after DNA replication (DeSombre et al.,
1988c). While at the lower doses there was a linear increase in chromosome breaks per cell with dose, at the high
dose there was such substantial chromosomal damage that it
was difficult to score the aberrations accurately (Fig 4.)

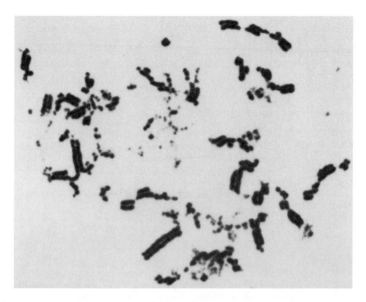

Figure 4. Effect of 80mBr-UdR on chromosome breakage. Metaphase CHO cells exposed to 26.7 µCi/ml ^{80m}Br-UdR for 2 hr followed by culture for an additional 10 hr before harvest and chromosomal analysis. From DeSombre et al., 1988c.

STUDY OF BROMINE-80m-LABELED ESTROGENS

We decided to study two types of bromoestrogens, steroidal and non-steroidal triphenylethylene-based ligands because of their significantly different ER and non-specific binding characteristics. Along the synthetic route to an hydroxy tamoxifen analog with halogen on the aliphatic double bond, we studied the 2-bromo-1,1-bis[4-hydroxyphenyl]-2-phenyl-ethylene (Br-BHPE, Fig 5.) and found that it was a powerful estrogen and had excellent affinity for the estrogen receptor (DeSombre, et al., 1988b). With two identical substituents on carbon 1 (unlike the antiestrogens with the aminoethoxy side chain on one phenyl ring and phenyl or 4-hydroxyphenyl on the other) the synthetic and biologic complications of cis-trans isomerism were eliminated in the Br-BHPE. We therefore synthesized and studied radiobromine-labeled Br-BHPE.

Bromine-80m was prepared by irradiating >99% enriched selenium-80 on a cooled aluminum target with a 20 µA colli-

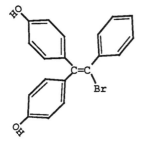

Figure 5.   Br-BHPE

mated 15 MeV proton beam using the CS-15 cyclotron at the University of Chicago, giving an end of bombardment yield of 1-2 mCi per μA-hr (DeSombre, et al., 1988a). The selenium target was dissolved and the radiobromine (a mixture of bromine-80m and the 17 minute daughter nuclide bromine-80) recovered. The [80mBr]Br-BHPE was prepared by reacting the bromine-80m with 1,1-bis (4-hydroxyphenyl)-2-phenyl-2-

tri-n-butylstannyl-ethylene in H_2O_2/glacial acetic acid, by the radiohalodestannylation reaction as reported by Hanson et al., 1984 (DeSombre, et al., 1988a). After a 10 minute reaction, the [80mBr]Br-BHPE was purified by reverse-phase high performance liquid chromatography. The simplicity of the preparation and purification of the [80mBr]Br-BHPE is important because of the short half life (4.4 hr) of the bromine-80m. In general we are able to obtain the final, purified product in less than one half life from the end of bombardment in the cyclotron.

As seen in Fig 6. sedimentation analysis of the binding of [80mBr]Br-BHPE to ER of rat uterine cytosol on low salt gradients demonstrates 8S binding, inhibitable by diethylstilbestrol (DES). The preparation shown gave a calculated specific activity of 8700 Ci/mmole, about 200 times the specific activity of the ^{3}H-estradiol. Fig. 7. confirms that the binding of [80mBr]Br-BHPE is actually to ER as it is recognized by the specific anti-ER antibody, H222.

When the [80mBr]Br-BHPE was administered to immature female rats and various tissues excised 0.5 and 2 hr later, the highest concentrations of bromine-80m were found in the uterus, Fig 8. The estrogen specificity of the tissue uptake of [80mBr]Br-BHPE was studied by co-administration of 1 μg of DES to some animals. As also seen in Fig 8, there was notable inhibition of the tissue content of bromine-80m in the ER+ target tissues, the uterus, ovary, pituitary and vagina. Although significant amounts of bromine-80m were also found in the intestine, the radiobromine levels in this tissue were no different in the presence and absence of DES. Furthermore, when the intestinal contents were assayed, most of the nuclide was found in the contents, rather than in the tissue itself as expected if the radioactiv-

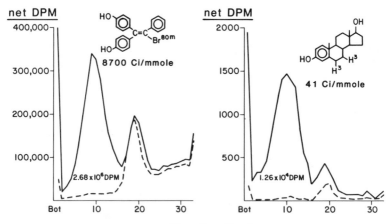

Figure 6. Sedimentation analysis of the binding of [80mBr]Br-BHPE to low salt extractable estrogen receptor from immature rat uterus. Saturating concentrations of [3H] E2 (41 Ci/mmole) and [80mBr]Br-BHPE, alone (solid lines) or in the presence of 1 μmolar DES (dashed lines), were incubated with the ER-containing extract, excess un-bound estrogen removed with DCC, the labeled extracts lay-ered on 10-30% sucrose in T10K10E1, pH 7.4, and centrifuged for 14 hr at 208,000 X g, 2°. Fractions were collected from the bottom of the gradients, counted in a γ or scin-tillation counter, corrected for counting efficiency and, for bromine-80m also corrected for decay. Bot indicates the bottom of the gradient. From DeSombre et al., 1988a

ity represents excretion. The liver was found to contain significant amounts of the radiobromine, but the levels in this tissue were also the same in the presence and absence of DES. While it has been reported that the liver has low levels of ER, it is well known that the liver metabolizes estrogens, and such extranuclear localization of most of the bromoestrogen would be compatible with the lack of DES inhibition of uptake by the liver. As seen in Fig 8, the pattern of tissue distribution was very similar at 0.5 and 2 hr, but with lower levels of nonspecific uptake (uptake in the presence of DES). At both time points there was a significant, DES-inhibitable uptake by the uterus, pitui-tary, and ovary, while the DES-inhibitable difference in the uptake by the vagina only was statistically significant at the 2 hr time point.

The synthesis of the 17α[80mBr]bromovinyl estradiol-17ß, [80mBr]Br VE$_2$, was carried out by a similar route with a tributyltin precursor (DeSombre, et al., 1988a). When ad-

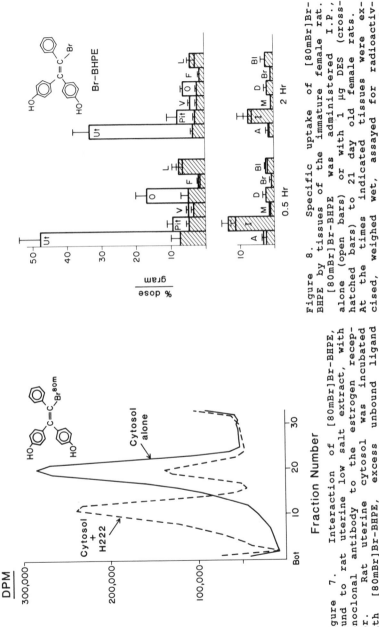

Figure 7. Interaction of [80mBr]Br-BHPE, bound to rat uterine low salt extract, with monoclonal antibody to the estrogen receptor. Rat uterine cytosol was incubated with [80mBr]Br-BHPE, excess unbound ligand removed with DCC, and the supernatant incubated with buffer (solid line) or 10 μg/ml monoclonal anti-ER antibody H222, and analyzed by sedimentation on sucrose gradients in the presence of 400 mM KCl as described in Fig 5. From DeSombre et al., 1988a

Figure 8. Specific uptake of [80mBr]Br-BHPE by tissues of the immature female rat. [80mBr]Br-BHPE was administered I.P., alone (open bars) or with 1 μg DES (cross-hatched bars) to 21 day old female rats. At the times indicated tissues were excised, weighed wet, assayed for radioactivity, the disintegration rate corrected for decay and related to injected amounts (ca.12-14 μCi/rat) of [80mBr]Br-BHPE (specific activity about 4000 Ci/mmole). From DeSombre et al., 1988a

ministered to immature female rats [^{80m}Br]Br VE2 also showed
significant, DES-inhibitable uptake by the classical estrogen target tissues, uterus, pituitary, vagina and ovary,
Fig 9. Although a relatively high uptake was seen at 0.5
hr in the liver (Fig 9), there was substantial variability
among the bromine-80m concentration of the liver, ovary,
adrenal and intestines at this time point as seen by the
large standard deviations. The variability was much lower
at the 2 hr time point when the specificity of uptake by
the uterus, pituitary, vagina and ovary were all clearly
evident. For all the other tissues studied there were no
differences in tissue concentration of bromine-80m when the
[^{80m}Br]Br VE$_2$ was administered alone or along with DES (Fig
9).

From a comparison of Figures 8 and 9, it appeared that
the uterine uptake of bromoestrogen was somewhat more favorable for the [^{80m}Br]Br-BHPE. On the other hand the radioactivity in the pituitaries after these two bromoestrogens were found to be more comparable. The particularly
high levels of radioactivity in the peritoneal target tis-

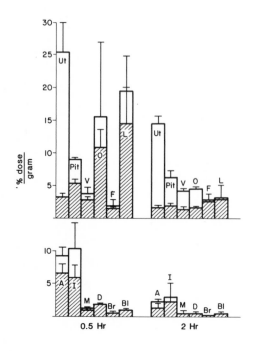

Figure 9. Specif-
ic uptake of
[^{80m}Br]Br-VE2 by
tissues of the im-
mature female rat.
[^{80m}Br]Br was ad-
ministered I.P.,
alone or with 1 µg
DES and analyzed
as indicated in
Fig 8. From De-
Sombre et al.,
1988a

sues, compared to that in the pituitary, are in part due to the I.P. route of administration When the tissue to blood ratios were compared for the [^{80m}Br]Br-BHPE and [^{80m}Br]Br VE$_2$ one of the significant differences between these two bromoestrogens became evident (DeSombre et al., 1988a). For [^{80m}Br]Br-BHPE, which shows more substantial binding to serum components, the bromine-80m content of blood showed more persistence than that following the administration of [^{80m}Br]Br VE$_2$. As a result the tissue to blood ratios for the ER target tissues tended to be higher for [^{80m}Br]Br VE$_2$ especially at the 2 hr time point. Further studies on the nature of the bromine-80m containing components in the blood will be necessary before one can conclude that the triphenylbromoestrogen, itself, has a longer half life in the blood, which might be advantageous for treatment purposes. However, it would appear that if diagnostic imaging were envisioned with γ-emitting haloestrogens, the steroidal estrogen type ligand might be preferred for the higher tissue to blood ratios for ER+ tissues.

The above studies indicate that both the steroidal and the nonsteroidal triphenylethylene bromoestrogens show favorable characteristics appropriate for use as estrogen receptor-directed therapy of ER+ cancers. The next stage of the work will involve preparation of larger amounts of the bromine-80m labeled estrogens at higher specific activities so that their cytoxicity can be tested in ER+ cells in vitro and in tumors in animals to provide quantitative information on the mean lethal dose of this nuclide when directed to bind near the DNA by virtue of the specific association of the estrogen with acceptor sites in DNA.

REFERENCES

Beatson, G.T (1896). On the treatment of inoperable cases of carcinoma of the mamma. Suggestions for a new method of treatment with illustrative cases. Lancet 2: 104-107.

Bloomer WD, McLaughlin WH, Weichselbaum RR, Tonnesen GL, Hellman S, Seitz DE, Hanson RW, Adelstein SJ, Rosner AL, Burstein NA, Nove JJ, Little JB (1980). Iodine-125-labelled tamoxifen is differentially cytotoxic to cells containing oestrogen receptors. Int J Radiat Biol 38: 197-202.

Bronzert DA, Hochberg RB, Lippman ME (1982). Specific cytotoxicity of 16α -[^{125}I] iodo-estradiol for estrogen receptor-containing breast cancer cells. Endocrinology 110: 2177-2179.

DeSombre ER, Carbone PP, Jensen EV, McGuire WL, Wells SA, Wittliff JL, Lipsett MB (1979). Steroid receptors in breast cancer. New Engl J Med 301: 1011-1012.

DeSombre ER, Holt JA, Herbst AL (1987). Steroid receptors in breast, uterine, and ovarian malignancy. In Gold JJ, and Josimovich JB, (eds): Gynecologic Endocrinology, Plenum, New York, pp 511-528.

DeSombre ER, Mease RC, Hughes A, Harper PV, DeJesus OT, Friedman AM (1988a). Bromine-80m-labeled estrogens: Auger electron-emitting, estrogen receptor-directed ligands with potential for therapy of estrogen receptor-positive cancers. Cancer Res 48: 899-906.

DeSombre ER, Mease RC, Sanghavi J, Singh T, Seevers RH, Hughes A (1988b). Estrogen receptor binding affinity and uterotrophic activity of triphenylhaloethylenes. J Steroid Biochem 29: 583-590.

DeSombre ER, Harper PV, Hughes A, Mease RC, Gatley SJ, DeJesus OT, Schwartz JL (1988c). Bromine-80m radiotoxicity and the potential for estrogen receptor-directed therapy with Auger electrons. Cancer Res 48: in press.

Everson RB, Hall TC, Wittliff JL (1973). Treatment in vivo of R3230AC carcinoma of the rat with estradiol mustard (NSC-112259) and its molecular components. Cancer Chemother Rept 57: 353-9.

Hanson RN, Seitz DE, Botarro JC (1982). E-17α-[^{125}I]iodovinylestradiol: An estrogen-receptor-seeking radiopharmaceutical. J Nucl Med 23: 431-436.

Hanson RN, Seitz DE, Bottaro JC (1984). Radiohalodestannylation: Synthesis of I-125 labeled 17α-E-iodovinylestradiol. Int J Appl Radiat Isot 35: 810-815.

Hochberg RB, and Rosner W (1980). Interaction of 16α-[^{125}I]iodo-estradiol with estrogen receptor and other steroid-binding proteins. Proc Natl Acad Sci USA 77: 328-332.

Horwitz KB, McGuire WL, Pearson OH (1975). Predicting response to endocrine therapy in human breast cancer : An hypothesis. Science 189: 726-727.

Jagoda EM, Gibson RE, Goodgold H, Ferreira N, Francis BE, Reba RC, Rzeszotarski WJ, Eckelman WC (1984). [I-125] 17α-iodovinyl 11ß-methoxyestradiol: In vivo and in vitro properties of a high-affinity estrogen-receptor radiopharmaceutical. J Nucl Med 25: 472-477.

Kassis AI, Adelstein SJ, and Bloomer WD (1987). Therapeutic Implications of Auger-emitting Radionuclides. In:

Spencer RP, Seevers R, Friedman A, (eds): Radionuclides in Therapy, CRC Press, Boca Raton, FL, pp 119-134 .

Katzenellenbogen JA, Senderoff SG, McElvany KD, O'Brien HA, Jr, and Welch MJ (1981). 16α[^{77}Br]Bromoestradiol-17ß: A high specific activity, γ-emitting tracer with uptake in rat uterus and induced mammary tumors. J Nucl Med 22: 42-47.

Katzenellenbogen JA, McElvany KD, Senderoff SG, Katzenellenbogen BS, and the Los Alamos Medical Radioisotope Group (1982). 16α[^{77}Br]Bromo-11ß-methoxyestradiol-17ß: A gamma-emitting estrogen imaging agent with high uptake and retention by target organs. J Nucl Med 23: 411-419.

McElvany KD, Carlson KE, Welch MJ, Senderoff SG, Katzenellenbogen JA, and the Los Alamos Medical Radioisotope Group (1982). In vivo comparison of 16α-[^{77}Br]bromoestradiol-17ß and 16α [^{125}I]iodoestradiol-17ß. J Nucl Med 23: 420-424.

McGuire WL, Carbone PP, and Vollmer EP (eds) (1975): Estrogen Receptors in Human Breast Cancer, Raven Press, New York.

Nardulli AM, and Katzenellenbogen BS (1986). Dynamics of estrogen receptor turnover in uterine cells in vitro and in uteri in vivo. Endocrinology 119: 2038-2046.

Powell GF, DeJesus,T, Harper PV, and Friedman AM (1987). A Monte Carlo treatment of the decay of ^{80m}Br, a novel Auger electron emitting isotope with potential for therapy. J Radioanal Nucl Chem Letters 119: 159-170.

Seevers RH, Mease RC, Friedman AM, and DeSombre ER (1986). The synthesis of non-steroidal estrogen receptor binding compounds labeled with Br-80m. Nucl Med Biol 13: 483-495.

INDEX